D1025312

STATISTICS YOU CAN'T TRUST

STATISTICS YOU CAN'T TRUST

A Friendly Guide To Clear Thinking About Statistics In Everyday Life

Steve Campbell, Ph.D.

Illustrated by Mark V. Hall

Making Critical Thinking Fun

Parker, Colorado

Published 1999 by Think Twice Publishing

Inquiries should be addressed to: Think Twice Publishing, Box 472, 8915 South Parker Road, Parker, CO 80134, Attention: Steve Campbell. Phone: (303) 627-9740. FAX: (303) 680-3863.

Library of Congress Catalog Card Number
98-90593
ISBN 0-9666171-5-0

1. Statistics 2. Critical thinking

10 9 8 7 6 5 4 3 2

Published in the USA

Many Thanks To ...

- Professors Beth Hennessey of Wellesly College, Bernard Grofman of the University of California, Irvine, and Thomas Obremski of the University of Denver for reading the entire book and making many incisive suggestions for improving it.

- Authors Morris James Slonim (*Sampling In A Nutshell)* and Darrell Huff (*How To Lie With Statistics, How to Take A Chance,* and others) for demonstrating long ago in their own writings that a pinch of humor can do wonders for the appeal of the subject of statistics.

- Cartoonist Mark V. Hall for his many off-beat ideas and his skill in translating them into illustrations the viewer can understand and enjoy.

- Kara King of Eagle Eye Editing for doing very well all the things an editor is expected to do, including, it seems, keeping the writer appropriately humble.

- Legions of university students who dutifully brought me over the years examples of misuses of statistics some of which appear in these pages.

- Several authors, editors, and advertisers who granted me permission to quote copyrighted material even though they must have suspected that I would be more critical than complimentary.

- My marvelous wife Judy for being tirelessly supportive.

...it was the age of wisdom, it was the age of foolishness...

—Charles Dickens

A Tale Of Two Cities

CONTENTS

INTRODUCTION

> Statistical thinking will one day be as necessary for efficient citizenship as the ability to read and write.
>
> —H. G. Wells

STATISTICS! Now there's an acquired taste if ever there was one. In a popularity contest "statistics" would probably place somewhere between "broccoli" and "root canal." But let us not forget that broccoli is good for us, root canals have spared many a bad tooth the anguish of extraction, and statistics are what we all use when we wish to communicate information and ideas in a precise way.

Actually, statistics is the most interesting of these three topics to read and write about because it is the most indispensable. That hasn't always been so; but it is now, as H. G. Wells (quoted above) foresaw with remarkable clarity nearly a century ago. Now, as we await the arrival of an exciting new century, we find ourselves living in a world where statistical facts, statistically laced arguments, and decisions based on statistics are all part of everyday life.

My only quarrel with Wells is that I think he should have said, "statistical *straight* thinking... ." rather than just "statistical thinking... ." Fuzzy or twisted thinking, whether statistical or otherwise, seldom benefits anyone. Regrettably, in this supposedly enlightened age of ours, an enormous amount of fuzzy and twisted thinking is served up in statistical form. Yes, and much valid statistical information gets interpreted in fuzzy and twisted ways. That might always be so despite my efforts to exorcise these evils. As I will illustrate shortly, I have found the twin devils of poor statistics and poor interpretation of good statistics to be pretty entrenched; they hold fast, yielding only to the rare individual with the skills and weapons to make them submit. Which brings us to you...

You can be one of those rare individuals. This book is your drill instructor and your armory. It will make you a better judge of statistical data, statistical graphics, and statistical arguments of a vast variety of kinds. It will help you see through statistical information not worthy of acceptance and certainly not worthy of being used as a basis for decision-making. It will also put you safely out of reach of the dangers and heavy costs endured by many as a result of fraudulent statistical claims. At the same time, it will make you feel more confident when you *do* accept a piece of statistical evidence, whether that evidence pertains to your business activities, the law, consumer protection, an important social issue, or practically anything else.

Who am I to be making such grand claims? I am someone who for many years has been a dedicated MSC (Master Statistical Critic, an accolade you will be entitled to bestow on yourself just as soon as you have finished reading this book and blown the whistle on at least one big shot—nothing lower than a congressman, please—found to be misusing statistics. In the meantime, your official title is Fledgling Statistical Critic, or FSC). I am also someone who wrote a rather successful book on this same subject back in the days of oil shortages, galloping inflation, and the

Watergate scandal. A brief tour of the earlier book, *called Flaws and Fallacies in Statistical Thinking* (hereafter just *F and F*), will give you a sense of what to expect from this one. It will also make you aware of some of my successes and failures:

First of all, the experience was, for me, mostly pleasant. Sales of *F and F*, I am told, were in excess of 50,000 and these were spread over at least seven countries. The book was printed in Japanese, Spanish, and (rumor has it) Chinese, as well as in English. Despite its advanced age (23 years), it remained in print and in use until a short time ago when I requested its euthanasia. Finally, and most important, readers have said mostly good things about it.

All in all, it would seem that I am entitled to take credit for helping a large number of gracious readers approach statistical evidence more wisely. That is gratifying indeed. Do I now bask in a feeling of self-importance? Hardly. For that to happen, I would have to see clear evidence that I have changed the world for the better. I would have to see some reduction in the total universe of bad statistical arguments. And I would have to see more people talking back to such arguments. As the following retrospective illustrates, I have learned that this world of ours isn't easily changed:

In *F and F*, I accused the Volvo Automobile Company of making questionable advertising claims regarding the durability of its cars. One of my complaints at the time had to do with their assertion that "...95% of all Volvos registered here in the past 11 years are still on the road." I smugly pointed out that approximately half of those sales had been made during the most recent four years and that the company's figures indicated nothing meaningful about the durability of Volvo automobiles. Did my mini-crusade accomplish anything? Well, as far as I know, Volvo did stop making that claim. However, if I can take any credit for the change—and that is not at all certain—I fear that the ensuing rapture didn't last very long: Between 1974, the year the earlier book was published by Prentice-Hall, and now, I have heard much the same claim made, in the same misleading way, on behalf of Ford, Chevrolet, and Dodge Trucks and several other makes of automobiles.

Anyway, Volvo went straight. Right? Not exactly. I wasn't terribly aware of Volvo's activities during the '80's, but looked on with interest in 1990 when the company got into trouble with the Federal Trade Commission over a blatantly misleading television commercial. Then, in late 1996, Volvo was "honored" with a Hubbard Lemon Award, bestowed by the Coalition For Science In The Public Interest for yet

another stinko television commercial. I really don't know what to think about those rascals. People tell me that Volvos are good cars; one would think the company could do very nicely just telling the truth. Anyway, do you see why I can't claim any glorious victories in this arena?

In *F and F*, I playfully injected a true historical nugget about demons: I reported that a German named Weirus, who lived in the latter part of the Sixteenth Century, claimed to know how many demons existed. He said there were exactly 7,405,926 demons and these, he insisted, were divided into 72 battalions, each under a prince or captain. (His data-gathering methods appear to have been proprietary.)

When I wrote that, I was naive enough to think that no one believes in demons anymore. Since then, thanks to the New Age movement, stories about satanic cults, a few sick rock bands, Shirley MacLaine's many best-sellers, numerous tabloid newspapers not at all concerned about the distinction between fact and fiction, slick new entries in the UFO-abduction sweepstakes, and some pretty pathetic television talk shows willing to say or do anything to titillate their gullible viewers, we are once again awash in baseless superstition. (Maybe it was always so and I just hadn't noticed.) Anyway, the world might have changed, but in a way which I am delighted to say I had nothing to do with.

In the earlier book, I called attention to the disconcerting habit of politicians of different allegiances to draw totally different conclusions from the exact same economic data. Sure enough, in the presidential debates of 1996, President Clinton proudly reminded his audience that, since he took over the presidential reigns in 1992, unemployment has declined from an unacceptable 7 percent rate to a much healthier 5.3 percent. His opponent, Robert Dole, on the other hand, lamented the fact that the so-called "healthier" rate masks the ill effects of widespread downsizing, one such ill effect being great numbers of people working at jobs beneath their training and abilities.

Neither speaker was really wrong—and neither was actually lying. It's just that neither was willing to present a very balanced account of the subject. Depend on it. Until the end of time, whatever the hot topic is, one politician will look at the numbers and say "good" and the other will look at them and say "bad." How can I hope to change the world when I can't even change the politicians?

An extremely magnanimous reader might be willing to give me credit for the downfall of the Soviet Union. At any rate, I did say in the earlier book that the Russian people, and many others around the world, had been lied to with statistics trumped up by Gosplan, the central Soviet planning agency, to make communism appear more conducive to rapid economic growth than capitalism. Alas, close friends have intimated that *F and F* was just one influencing factor among many leading to the Soviet demise—and not necessarily the most important one.

Clearly, if changing the world is the appropriate criterion, the evidence on my side is scant indeed. But I am not discouraged. If I have learned anything from being a parent, it is this: when confronted by evidence that one has been an abject failure, there is only one constructive thing to do: Yell. The reader of both books might find my tone somewhat more militant this time around.

My growing crankiness aside, and overlooking the fact that I have downgraded the goal of changing the world to a mere wish to do so, the aim of *Statistics You Can't Trust* is the same as its predecessor, namely to make available a nontechnical, hopefully entertaining, book beneficial to two groups of people. The first group consists of university students taking a first course in statistical methods. Although a growing number of statistics textbooks now include some material on ethics, most stick pretty close to the whens and hows of formal statistical methodology. These, textbooks, already very large, seldom have much content aimed at helping the student to better distinguish between valid and bogus statistical claims.

Consequently, *Statistics You Can't Trust* is offered as a supplemental reading text to help the captive statistics student become a better informed, more critically minded, citizen. The fact that the sequence of topics is in rough accord with that of many beginning statistics textbooks will, I trust, prove an advantage. (Naturally, it would warm my golden years to learn that my lovechild had also been put to good use in courses other than statistics. Logic, journalism, debate, and law spring to mind.)

The second group I hope to reach is made up of ordinary citizens, many of whom will have had no formal statistical training at all. We are all bombarded by questionable claims from politicians, lawyers, advertisers, TV magazine show hosts, journalists, salespeople, various kinds of experts, and outright charlatans practically every day of our lives. Add to this the fact that, where numbers are concerned, we are all susceptible to much self deception, and you have what might euphemistically be called "a situation in need of repair." This book attempts to make a contribution to the repair work.

Hopefully, it will be used as a self-help guide for the statistical novice who feels the need to evaluate claims and evidence more judiciously. Toward this end, the terminology used and the manner in which the topics are treated are based on the assumption that the reader has had no classroom exposure to the subject of statistics and that his or her background in mathematics is limited to basic arithmetic. On the other hand, since some of the material is tightly reasoned, another assumption is that the reader is intelligent and sufficiently tenacious to wrestle with such resistent material until it is properly subdued.

Just a brief word about the examples in *Statistics You Can't Trust,* and we will get on with the book. You will quickly find that most of the examples are good examples of bad statistics (not to be confused with bad examples of good statistics). That should not be interpreted as implying that statistical evidence is usually wrong. Rather it is a reflection of my agreement with the views of Ernst Wagemann (in *A Fool's Mirror of Statistics*): "We share with Socrates the pious hope that men avoid mistakes once they are aware of them. But we are frivolous enough to suppose that men do this out of a spirit of pure contrariness, and hence are more affected by the sight of a horrible example than a good precept."*

Onward, Fledgling Statistical Critics! You have nothing to lose but your gullibility.

*Translated from the German by Norma E. Kruskal.

1

THE LAST SHALL BE FIRST

> ...there is in our time no escape from figures... It is not only that again and again the best evidence is quantitative, although this is both true and important. It is also that many of the most disputed contemporary issues are themselves essentially quantitative.
>
> —Antony Flew*

It won't do for you to say you are not interested in statistics. You might as well profess a disinterest in life itself. The two—statistics and life—have come to be pretty intertwined.

Whatever your feelings about statistics as a field of study, I think you will agree that it is today very difficult to arrive at an enlightened position on almost any important issue without first being confronted by, if not bombarded by, other people's statistics. Whether the subject is the environment, drugs, crime, or economic conditions, the press and spokespersons for all sides will usually offer up some statistics to enhance their positions. And why not? *Good* statistical evidence rightfully commands more respect than mere impression or conjecture.

**Thinking Straight,* (Amherst, NY: Prometheus Books, 1977), p. 85.

Of course, statistical information is used for much more than just bolstering the respectability of a point. Through statistics, modern man evaluates business activity; records social progress; elects presidents and keeps abreast of their popularity; measures intelligence, interests, and aptitudes; determines which television shows will survive and which will not; compares the profit potential of alternative business strategies; decides whether to invest in stocks or bonds and whether now is a good time to buy; keeps track of batting averages, yards per carry, and field goal percents; and in a vast variety of other ways, keeps informed about what is going on in the world.

Regrettably, although the quality of some statistical evidence is unimpeachable, far too much is worthless. Some is downright misleading. Quite a lot is bad because the hands it passes through belong to well-meaning but statistically unsophisticated people. For example, the journalists I have known are not fiendish sociopaths; most wish to impart correct information to the rest of us.

Unfortunately, few such journalists have had any more training than their readers or viewers in subjects leading to skill in evaluating statistical evidence. I mention this in case you harbor the same misconception I did for many years, namely that the press meticulously filters out the misinformation and passes only valid information along to the public. The public—you and I—must do its own filtering.

Perhaps the most efficient way to illustrate the near ubiquity of statistics in today's world and the desirability of approaching statistically based arguments with a pinch of caution would be to call your attention to the diagnostic quiz immediately below. This quiz will also give you a chance to see what kind of statistical eye you already have. Give it a serious try. (But realize that, whatever your score, I, like any other self-respecting author with a diagnostic quiz in his pack, will use it to argue that your need for this book borders on desperate.) For each of the following statements, if the statement sounds more true than false, jot down "True" on a piece of paper. If the statement sounds more false than true, jot down "False."

1. I appreciate The National Safety Council's information that most automobile fatalities occur at speeds of under 40 miles an hour because now I know that, for safety's sake, I should always drive like a bat out of hell. **True or False?**

2. Soldiers fighting in a place where bombs are falling should climb into fresh shell holes because it is highly unlikely that a second bomb will hit in exactly the same spot. **True or False?**

3. If I were to learn that ninety percent of heroin users had, before developing the heroin habit, used marijuana regularly, I could properly conclude that marijuana use leads to heroin use. **True or False?**

4. If one of my salesmen had told me a year ago, "I sold $100,000 worth of merchandise this year and intend to sell $110,000 next year," I would be quite favorably impressed if he *now* told me, "I sold $102,000 worth of merchandise, which means that I achieved ($102,000/$110,000)•100 = 92.7% of my goal." **True or False?**

5. If my uncle had been suffering from severe arthritis pain for years and then tried a new experimental drug, I would have to give the new drug credit if the pain went away. **True or False?**

6. If I were to learn that ninety percent of the murders involving guns in this country were committed by normally law-abiding citizens, either by accident or in a moment of great emotional trauma, I would tend to support stricter gun-control laws in the expectation of reducing the number of such unfortunate occurrences. **True or False?**

7. I believe that police departments around the world have benefitted immeasurably from the help of psychics who are able to sense where a body is located and what the killer looks like. **True or False?**

8. The decade of the eighties has the dubious distinction of showing a faster percentage growth in emotional immaturity among U.S. citizens than any other decade of the Twentieth Century. **True or False?**

9. Satanists are responsible for as many as 50,000 human sacrifices a year, mainly transients, runaways, and babies conceived solely for the purpose of human sacrifice. **True or False?**

10. If I were to learn that a political polling service had questioned several thousand potential voters and found that 70 percent favored the Republican candidate for President of the United States, I would conclude that the Republican candidate was bound to win. **True or False?**

11. If in a certain school district in 1990, 25 percent of the children belonged to single-parent families and in that same school district in 1995, 28 percent of the children belonged to single-parent families, the increase between 1990 and 1995 would be three percent. **True or False?**

12. I am in Paris, France for the first time in my life where I run into an old school chum who is also here for the first time, I learn that we both have sweethearts named Pat and that we both have had a parent die during the past year, I return to my hotel room, Number 456, only to find that my chum is staying in the same hotel and is occupying Room 654. I conclude that I have been a participant in an extraordinary coincidence. **True or false?**

Let us examine each statement briefly. I hope you jotted down a lot of "Falses" because I view all of these statements to be false or, at least, more false than true.

1. I appreciate the National Safety Council's information that most automobile fatalities occur at speeds of under 40 miles an hour because now I know that, for safety's sake, I should always drive like a bat out of hell.

False. The Council's point was, in substance, "Wear seat belts even if you drive slowly." That, of course, is sound advice. But think about how easily advice like that can backfire: For example, suppose I tell you that far more people die while sleeping than while bungee jumping? That is undoubtedly true because we all sleep and only a few of us bungee jump. Still, I hope you wouldn't conclude that, for safety's sake, you should exit your bed immediately and seek solace on the nearest bungee-jumping platform.

These examples are humorous—even silly. But ponder for a moment the potential treachery hidden behind the humor. For example, if I am able to convince you to go bungee jumping by assuring you that it is safer than lying in bed, I will have misled you in a way that has potentially grave (so to speak) consequences. If I can persuade you that smoking cigarettes is safer than crossing the street, I will have done you another disservice. And don't let anyone convince you that flying in an airplane is safer than taking a bath or that being shot out of a cannon is safer than getting a manicure. Granted, there might conceivably be a valid basis

for each of these preposterous-sounding claims, but unless you are well acquainted with the underlying data and the reasoning employed, you will be wise to assume that the choice which sounds safer *is* safer.

I have dubbed this fallacy the Risk-Your-Life-and-Live-Longer fallacy. It is treated further in Chapter 7.

2. Soldiers fighting in a place where bombs are falling should climb into fresh shell holes because it is highly unlikely that a second bomb will hit in exactly the same spot.

False. If, before the fact, we were to ask, "What is the probability that a bomb will hit in the exact same designated spot twice during a short period of time?," we would undoubtedly conclude that the probability would indeed be very small. However, once the bomb has hit the designated spot, it is just as likely to hit there again as it is to hit any other specified spot of the same size in the same general vicinity. Support for this claim lies within the realm of probability theory, a subject treated in Chapter 9.

By the way, this is not just something stupid I made up to enliven a book. American soldiers during World War I were actually given this advice. Thus, a statistical fallacy is seen to have been an aspect of battlefield conduct. Whether any harm resulted from it, I can't say.; however, It certainly must have led to a false sense of invincibility.

3. If I were to learn that ninety percent of heroin users had, before developing the heroin habit, used marijuana regularly, I could properly conclude that marijuana use leads to heroin use.

False. At least this is false in the sense that only one half of the story is being told. Imagine, if you will, three groups of people, A, B, and C. Group A is very large and is made up of those people who use marijuana regularly. Group B is much smaller and is made up of those people who use heroin regularly. Finally, Group C is slightly smaller than Group B and is made up of those people who previously used marijuana regularly and then went on to use heroin regularly. Remember: Group A is large, Group B is smaller and Group C is smaller still.

Now, if we divide the number of people in Group C by the number of people in Group B (and multiply by 100), we have the percent of heroin users who used to be marijuana users. That is how the "ninety percent" in Statement 3 was figured. The fact that Group C is a large percent of Group B may seem impressive. However, does it really convey any useful information? Maybe so. But let us withhold judgment until we have figured the other possible percent.

If we divide the number in Group C by the number in Group A (and multiply by 100), we get a very small value. In words: We find that only a small percent of those who use marijuana regularly go on to use heroin regularly—and that is really what Statement 3 is about. That is, the statement has to do with what marijuana use maybe leads to—not with what maybe led to heroin use, a subtle distinction to be sure.

This fallacy is what I call the Opportunistic Construction of a Percent (OCP), a subject treated more completely in Chapter 6. OCP is one of the most treacherous of the many kinds of fallacies you will encounter as you make your way through this book—and through life. Characteristically, one special-interest group seizes upon one way of computing the percent and another special-interest group seizes upon the other way. Each group then attempts to beat the other into submission using its own favored percent. The substantive aspects of the related arguments are usually highly emotion-charged, socially important, and are occasionally matters of life or death. At the same time, the statistical aspects are often fiendishly subtle.

4. If one of my salesmen had told me a year ago, "I sold $100,000 of merchandise this year and intend to sell $110,000 next year," I would be quite favorably impressed if he *now* told me, "I sold $102,000 worth of merchandise, which means that I achieved ($102,000/$110,000)•100 = 92.7 percent of my goal."

False. Be favorably impressed if you like, but it's a pretty strange way to measure achievement. If this salesman had not increased his sales at all during the recent year, he still would have (according to his way of doing the calculations) achieved ($100,000/$110,000)•100 = 90.9% of the goal. If sales had dropped by $10,000, rather than increasing by the same amount, he would still be able to claim ($90,000/$110,000)•100 = 81.8% achievement of his goal.

A better idea is to take the realized increase, $2000, and divide it by the planned increase, $10,000, (and multiply by 100). That would give ($2000/$10,000)•100 = 20%.

5. If my uncle had been suffering from severe arthritis pain for many years and then tried a new experimental drug, I would have to give the new drug credit if the pain went away.

False. Oh yes, the new drug might have been just what your uncle needed. To demonstrate my open-mindedness, I'll let you claim half credit if you answered "True."

Nevertheless, there is a very important point to be made here:

Whenever we read causality into a situation involving events separated in time, we run the risk of calling the wrong thing the cause. Doing so can lead to serious, sometimes fatal, consequences.

For example, around the turn of the century a man named W. G. Brownson "invented" a therapeutic ring—the Electro-Chemical ring by name. Except for Brownson's grandiose claims, there was nothing very special about the ring; it was made of nothing more than commercial-grade iron!

Still, Brownson advertised that when worn on a finger regularly, the ring cured diseases caused by acid in the blood. And what, you ask, are some diseases caused by acid in the blood? Merely Bright's Disease, diabetes, epilepsy, varicose veins, cataracts, and the list goes on until fully 21 of mankind's more serious illnesses are mentioned.

Was Brownson jailed or fined? Neither. In fact, Until the *American Medical Association* called his charlatanry to the public's attention, Brownson did a fine business. Moreover—and here is where such stories take on a kind of Twilight Zone quality—sales were aided by the freewill testimonials of sufferers of most of the ailments on Brownson's list of 21! The people involved were, for the most part, honest folks who really thought they had benefitted from wearing the cure-all ring.

So what, you ask? If these decent people felt better, didn't the phony cure-all do some good? Doubtful. Many of the ailments on the list were of a serious, even life-threatening, nature. If some of the sufferers could really have been helped by conventional medicine, but no longer felt a need for such help, then the net effect of the ring was hardly good.

At the time a quack cure seems to be working, It is ballyhooed in a variety of ways. That is the time when the well-meaning sufferers volunteer testimonials—only to die shortly thereafter. Of course, the public usually only knows about the testimonials; the deaths merit only a few lines in the victims' hometown newspapers.

The task of determining whether a new medical treatment is really the cause of a patient's subsequent physical condition is fraught with complexity. We will return to this subject, and to the more general subject of identifying causes, in Chapter 11.

6. If I were to learn that ninety percent of the murders involving guns in this country were committed by normally law-abiding citizens, either by accident or in a moment of great emotional trauma, I would tend to support stricter gun-control laws in the expectation of reducing the number of such unfortunate occurrences.

False. This is much like Statement 3 above. Imagine three groups of people, A, B, and C. Group A is very large and is made up of gun owners. Group B is much smaller and is made up of those who have committed murder using a gun. Group C is smaller yet and is made up of normally law-abiding citizens who, nonetheless, have committed murder using a gun. Now follow the same procedure as in

I'M CURED DOC. THANKS TO YOUR MAGIC BEADS.

⇒

I TOLD HIM
TO HOLD BEADS
TIGHTER.
THUD

Statement 3. You should end up with two very different perspectives on the matter.

We will do this exercise together in Chapter 6. Also, a related subject, conditional probability, is treated in Chapter 13. (Notice that, once again, we find the fallacy of the Opportunistic Construction of a Percent associated with an issue where emotions run very high.)

7. I believe that police departments around the world have benefitted immeasurably from the help of psychics who are able to sense where a body is located and what the killer looks like.

False. Some people do indeed claim to receive psychic impressions about the way a crime was committed, why it was committed, where the body is located, and so forth. Unfortunately, the idea is sufficiently titillating to newspaper readers and television viewers that the media, again and again, have dealt with the subject as if it were an established fact rather than just a possibility. ("Cops Amazed At Crime-Busting Psychics" and "Can Psychics See What Detectives Can't" are just two actual titles of news articles on the subject.)

Moreover, fiction writers of various stripes have used the concept as a basis for books, magazine articles, and movies. The upshot is that many members of the public do really believe that "psychic detectives" have solved crimes by paranormal means which could not possibly have been solved by ordinary police-department grunt work.

Our motto should be "Extraordinary claims demand extraordinary supporting evidence." At this writing, the evidence, both statistical and qualitative, on behalf of the psychic detectives, is not very encouraging.* The limited statistical evidence we have suggests that you and I, if we had an abundance of audacity and minds that remembered little successes better than big failures, could do as well as the "real" psychic detectives. The topic is revisited in Chapter 12.

8. The decade of the eighties has the dubious distinction of showing a faster percentage growth in emotional immaturity among U.S. citizens than any other decade of the Twentieth Century.

False. This is what I call a *meaningless statistic.* It sounds for all the world as if someone has studied the subject of "emotional immaturity," decade-by-decade, and calculated and compared percent changes. Perhaps someone has done just that. If so, then, hopefully, the term "emotional immaturity" has been carefully defined. What remains is for the creator of the definition to share it with the rest of us. Such a clarifying definition is called an *operational definition*, a subject which, along with *meaningless statistic*, we return to in Chapter 2.

9. Satanists are responsible for as many as 50,000 human sacrifices a year, mainly transients, runaways, and babies conceived solely for the purpose of human sacrifice.

False. This claim comes from a deputy sheriff who was quoted in the *Kansas City Times.*** The trouble with this estimate is that it is apparently nothing but a guess.

*See Resier, et al, *Journal of Police Science and Administration,* March 1979.

** M. Berg, "Satanic Crimes Increasing? Police, Therapists Alarmed," March 26, 1988.

Granted, the deputy sheriff in this case might have a better feel for satanic-cult matters than we would. Still, the impression one gets from perusing the newspapers and magazines is that one can come up with just about any number one wishes.

For that matter, how many people do you suppose *belong to* satanic cults in the United States? Research seeking an answer found that estimates range between 150,000 and ten million—quite a range.*

This is an example of what I call an *unknowable statistic.* Nothing in life is perfect; just because we would like to know a statistical fact, doesn't necessarily mean that we ever will. What is important to realize is that often no one else has access to that information either. That being so, some statistical information is false on its face. We will explore *unknowable statistics* in Chapter 2.

10. If I were to learn that a political polling service had questioned several thousand potential voters and found that 70 percent favored the Republican candidate for President of the United States, I would conclude that the Republican candidate was bound to win.

False. There are two concerns here: First, realize that the presidential candidate preferred at Time A is not necessarily the one preferred at Time B—no small consideration if Time B happens to be election day. The polls mostly show how potential voters are feeling at a specific point in time.

*Paul D. Hicks, "Police Pursuit of Satanic Crime, Part II," *Skeptical Inquirer,* Summer, 1990, p. 379.

Second, the manner in which the sample was selected is of paramount importance. The sample must be representative of the relevant population of voters. If, for example, all the voters questioned resided in, say, Utah, historically a strong Republican state, we would have to declare that sample "unrepresentative." Under such circumstances, despite the large sample size, the collective opinions of the respondents are probably not indicative of the preferences of the nation's voters. The same conclusion would be appropriate if we were to learn that the sample was *self-selected.* Sampling is treated further in Chapter 10.

11. If in a certain school district in 1990, 25 percent of the children belonged to single-parent families and in that same school district in 1995, 28 percent of the children belonged to single-parent families, the increase between 1990 and 1995 would be 3 percent.

False. The 3 percent referred to resulted from taking the difference between the two percents given, that is, 28% - 25% = 3. Such a difference is properly called "3 percent points of change."

Assuming no change in number of families, the actual percent change could be computed by [(28 - 25)/25]•100 = 12 percent. This error, though seldom staggering in its consequences, is made by people in the news media quite often. The budding statistical critic really needs to know the difference between "percent" and "percent points." We return to this subject in Chapter 6.

12. I am in Paris, France for the first time in my life where I run into an old school chum who is also here for the first time, I learn that we both have sweethearts named Pat and that we both have had a parent die

during the past year, I return to my hotel room, Number 456, only to learn that my chum is staying in the same hotel and is occupying Room 654. I conclude that I have been a participant in an extraordinary coincidence.

False. Call it a coincidence, if you like. But I wouldn't get too carried away with the "extraordinary" part. We all have experiences just as "extraordinary" virtually every day (some would say constantly), but only occasionally are we sufficiently "startled" by them that we stop and reflect and tell others. To be sure, finding oneself the central figure in a coincidence can cause chills to run up and down the spine. It is difficult to shake the feeling that the experience was orchestrated by some kind of higher power.

Just ask Freedom A. Hunter: Hunter is the unfortunate teenager who somehow got the driver's license of one Tim Holt. Hunter also somehow got a checkbook belonging to a couple living near his home. Realizing that the two items together were rife with possibilities, Hunter proceeded to write a check for $275 payable to Tim Holt. However, when he went to a drive-up bank window to cash the check, using Holt's driver's license as identification, he soon found himself flanked by police cars. It seems that the bank teller assisting Hunter was none other than Tim Holt himself.*

Experiences like this notwithstanding, there is usually less to a coincidence than meets the eye. What makes coincidences potentially treacherous is their ability to help perpetuate such activities as the aforementioned psychic sleuthing and medical quackery. We return to coincidences in Chapter 9.

*Associated Press, February 2, 1991.

Score yourself according to the number of "False" responses—the more the better. Then forget the score as quickly as possible. Whatever your score, the quiz should have accomplished the following things:

- Provided an indication of whether your reactions to statistical claims are already on the critical side or whether they can use some nudging in that direction.

- Suggested that there are rather few facets of our lives completely unaffected by statistics.

- Demonstrated that many of the issues for which statistical evidence is brought to bear are matters of some gravity, pertaining as they do to our health, the national welfare, the safety of our citizens, the leaders we elect, and so forth.

- Shown, albeit without fanfare, that the word "statistics" itself has a longer reach than many suppose. Five of the 12 statements contain no actual numbers; nevertheless, they are still statistical in nature.

- Provided a glimpse of the kinds of topics to be treated in this book. Indeed, this chapter could easily have been the final chapter rather than the first. But we'll call it the first.

2

MEASUREMENTS: VALID, VEILED, AND OVERBLOWN

> Quantify. If whatever it is you're explaining has some measure, some numerical quantity attached to it, you'll be much better able to discriminate among competing hypotheses. What is vague and qualitative is open to many explanations.
>
> —Carl Sagan*

What is or are statistics? The word has two widely used meanings. The most generally familiar—and for many the least interesting—can probably be introduced most painlessly by the following excerpt from O. Henry's Handbook of Hyman:

> "Let us sit on this log by the roadside," says I, "and forget the inhumanity and ribaldry of the poets. It is in the columns of ascertained facts and legalized measures that beauty is to be found. In this very log we sit upon, Mrs. Sampson,

The Demon-Haunted World: Science as a Candle in the Dark (New York: Ballantine Books, 1996), p. 211.

> says I, "is statistics more wonderful than any poem. The rings show it is sixty years old. At the depth of two thousand feet it would become coal in three thousand years. The deepest coal mine in the world is at Killingworth near New Castle. A box four feet long, three feet wide, and two feet eight inches deep will hold one ton of coal. If an artery is cut compress it above the wound. A man's leg contains thirty bones. The Tower of London was burned in 1841."
>
> "Go on, Mr. Pratt," says Mrs. Sampson, "Them ideas is so original and soothing. I think statistics are just as lovely as they can be."

Although not all of Mr. Pratt's original and soothing ideas are really statistics, enough of them are to convey the idea that a statistic is a numerical fact such as a measurement, (the seven-foot height of your basketball team's center), a count (the twelve bulls-eyes achieved by Linda in an archery contest), or a rank (Dan's third-place standing in a bodybuilding competition). A statistic in this first sense can even be a summary measure such as a total, an average, or a percent.

Besides referring to numerical facts, the term "statistics" also applies to the broad discipline of statistical manipulation in much the same way that "accounting" applies to the entering and balancing of accounts. "Statistics" in this broader sense is a set of methods for obtaining, organizing, summarizing, presenting, and analyzing numerical facts.

Usually, these numerical facts represent partial, rather than complete, information about a situation, such as when a sample is used in preference to a more costly census. Subject to the following guidelines, the word will be used in both senses within this book:

The first eight chapters are primarily concerned with statistics in the sense of numerical facts. The term *descriptive statistics* will frequently be used in connection with these topics.

Later chapters provide a glimpse at statistics as an approach to making the most of partial information. Here, the terms *statistical inference* or *inductive statistics* will often be used.

The context within which the word "statistics" is found should make the intended meaning clear.

WHY ANCIENT LIARS PRAYED FOR STATISTICS

Since this book is about misuses of statistics, it would be nice if I could give you a brief history of such transgressions. Regrettably, neither I nor anyone else can do that very well.

We don't even know very much about when valid statistics first appeared. We do know that the first written records contain numbers, a fact suggesting that the ability to count goes way back. Fragmentary evidence suggests that the earliest attempts at census-taking might have been made around 4500 B.C. in Babylon. The Bible tells us that statistics in the purely "numerical facts" sense were used to provide information about taxes, wars, agriculture, and even athletic events.

Despite such tiny peeks at statistical usage in ancient times, we must surmise that there probably was a time when counting, and therefore statistics, was unknown, a time when a shepherd, for example, did not describe his flock as consisting of twenty, fifty, or one hundred sheep, but rather kept track of his woolly charges by assigning each a name. If two sheep turned up missing, the shepherd searched not for two anonymous animals, but for, say, Peter and Paul.

It seems likely that statistical fallacies—intentional or unintentional—first appeared about the same time as valid statistics. We all know people whose honesty we have good reason to doubt as well as people who are just plain careless or stupid. Certainly, there must have been people like that in the very beginning just as there must have been people like you and me—honest, meticulous, and worthy of becoming Master Statistical Critics had the honor been attainable in those primitive times.

Not much imagination is required to envision our shepherd, recounting the challenges he faced while retrieving his two wayward sheep, Peter and Paul, and rather than spoiling a good story by underselling it, claiming that he also had to search for Mary and Esther. The advent of counting and statistics certainly didn't create the all-too-human tendencies to lie, exaggerate, or make honest mistakes, but it did introduce a whole new, very colorful, means of giving vent to such tendencies.

THE ROOTS OF MEASUREMENT

In either of its two meanings, statistics is intimately tied in with the problem of measurement—the use of numbers to represent properties. Before one can study something scientifically, she must be able to express it in numbers, for only then can she distinguish easily and minutely between different but similar properties.

Unfortunately, the ideal way of expressing a property as a number may not be self-evident. Or if it is self-evident,

the physical procedures required may be prohibitively expensive or in some other way impractical to use. Researchers and professional data gatherers, therefore, must often resort to second- or third-best ways to measure whatever interests them. The procedures adopted determine, to a considerable degree, the validity of the data and the precise manner in which we, as consumers and evaluators of statistics, interpret them.

In the next several paragraphs we examine problems related to the task of measuring things. Because it is impossible to measure something meaningfully without first knowing what that something is, we must begin by concentrating on the crucial subject of definitions.

Sometimes the task of measuring a property is quite simple. Determining the weight of a sack of potatoes, for example, is for most of us a task of less than staggering proportions. If someone—let us say a philosophy student—were to ask, "Exactly how are you defining 'sack of potatoes'?" or "Are you planning to use avoirdupois weight or troy weight?" or "Can you prove the scales are accurate?" we might be moved to respond with something abrasive to the philosophy student's finer sensibilities.

On the other hand, when the thing being measured is "unemployment," "poverty," "marital compatibility," "mental health," "political popularity," or some other concept lending itself to many interpretations, the kind of cautious inquisitiveness displayed by our hypothetical philosophy student becomes absolutely essential. What could be less informative than a collection of figures purporting to measure, say, unemployment, when we are not even certain what kinds of people are counted as unemployed?

Are nonworking children included in the count? Are housewives? Self-employed professionals? Part-time employees? People on temporary layoff? In most cases, it probably matters less how such problems are handled—provided, of course, that they are handled sensibly—than whether we are told or left to guess about which categories of people are counted among the unemployed and which are not.

Whenever a term can be defined in a variety of ways, the data gatherer must decide which of the possible definitions seems best, and, just as important, which definition lends itself best to efficient, relatively inexpensive data collection. As a result, the definitions used are *operational definitions*, a term meaning that one of several definitions has been settled upon and the data user is asked to accept that specific definition when interpreting the figures. In return for this acceptance, the supplier promises to adhere religiously to that definition.

Why is an operational definition so important? There are two reasons: First, it gives solidity to a concept which might otherwise be pretty squishy. Second, the definition often has a great deal to do with the size of the related numerical fact or facts. The latter point can be illustrated rather dramatically through a now-classic example having to do with unemployment during the Great Depression:

How many people do you suppose were unemployed during, say, November of 1935? The answer is 9 million. Or 11 million. Or 14 million. Or 17 million. It depends on whose figures you like. According to Jerome B. Cohen, the bewildering and contradictory list of unemployment estimates shown on the following page all pertain to this one month and all were prepared by reputable agencies.

To make matters worse, just six months later, in May of 1936, the United States Chamber of Commerce issued a report estimating the number unemployed at 4 million, and the New York Sun announced that on the basis of a survey of 30 million workers, unemployment amounted to between 3 and 3.5 million. The Labor Research Association insisted that all estimates lower than its own were erroneous. The Chamber of Commerce held that all estimates higher than its own were inaccurate.

Table 2-1. Estimates of Unemployment for the Month of November, 1935, According to Five Reporting Agencies

Agency Preparing Estimate	Estimate Of Number Unemployed
National Industrial Conference Board	9,177,000
Govt. Committee On Economic Security	10,913,000
American Federation Of Labor	10,077,000
National Research League	14,173,000
Labor Research Association	17,029,000

Source: Jerome B. Cohen, "The Misuse of Statistics," *Journal of the American Statistical Association.* XXXIII, No. 204, (1938), 657.

These estimates, as you are aware by now, differ primarily because of differences in the definition of unemployment used. Some estimates took into account unemployment among farm labor and some did not; some included estimates of people leaving school and seeking employment for the first time and some did not. And so it goes...the considerable variation among the estimates testifying to the sometimes considerable *statistical leverage* exerted by a difference in definition.

As far as this country is concerned, such differences of opinion have been eliminated—officially at least. The U.S. Bureau of Labor Statistics releases unemployment statistics each month that are conceptually the most all-embracing in the world. Just about any age-eligible person who can possibly be construed as unemployed is, in theory at least, included in the count.

As long as we are aware that the operational definition of unemployment in this country is so all-embracing, we can use the figures in many meaningful ways. But problems still arise when making comparisons between countries. Many foreign unemployment estimates are based on registrations at employment exchanges, an approach tending to produce smaller unemployment rates than those for the United States.

THE DEFINITION THAT WASN'T THERE

The late columnist/humorist Robert Benchley once revealed to the world the following astonishing "fact: " "It is not generally known, I believe, that one comic editor dies every 18 minutes, or, at any rate, feels simply awful." This, of course, is a put-on. However, if you will keep a close eye on the news media, you will find a lot of this sort of thing reported with utmost seriousness: "Seventy-two percent of high school graduates have mixed feelings about college." What, pray, is the precise meaning of "mixed feelings?"

"Thirty-seven percent of New Yorkers are mentally ill, a rate that surpasses that of any other major city in the world." That's awful; but before you go any further, tell me what you mean by "mental illness" and "major city."

"During the past thirteen years there has been a one-hundred-percent increase in the number of children in our grade schools who are hyperactive." A cause for concern, yes. But, first, tell me the criteria used to declare a child "hyperactive." *

A top executive of a direct sales company was quoted as saying, "About 70 percent of the entire organization has been with the company for many, many years." Impressive. But how many is "many, many?"

In Chapter 1 you were asked to evaluate this statement: "The decade of the eighties has the dubious distinction of showing a faster percent growth in emotional immaturity among U.S. citizens than any other decade of the Twentieth Century." Do you see this statement in a different light now? Or were you already aware of its emptiness?

The above are all examples of *meaningless statistics*. A meaningless statistic is a precise figure used in conjunction with a term, or terms, sufficiently vague that an operational definition is sorely needed to endow the figure with meaning. But no such operational definition is provided.

*It isn't as easy as it might sound. See Diane McGuinnes, "Attention Deficit Disorders: The Emperor's Clothes, Animal 'Pharm,' and Other Fiction," ed. S. Fisher and R. Greenberg, *The Limits of Biological Treatments for Psychological Distress* (Hinsdale, New Jersey: Lawrence Erlbaum Associates, Publishers, 1989), pp. 151-187.

THE HEARTBREAK OF HYPERACCURACY

The story is told about a man who, when asked the age of a certain river, replied that it is 3,000,004 years old. When asked how he could give such accurate information, his answer was that four years ago the river's age was given as three million years.

Clearly, the man in this story was unaware that the three-million figure was a crude estimate rather than a precisely known fact. His tacking on the four years was not only unnecessary but potentially misleading as well, for it gave the impression that a degree of accuracy had been achieved that was really unattainable. This is an example of what I call *hyperaccuracy.* Some things simply cannot be measured with as much accuracy as some purveyors of statistical information like to pretend.

An automobile advertisement caught the reader's attention with the assertion that on a certain day 262,825,033.74 tons of snow fell upon the United States.

The official publication of the Austrian Finance Administration once stated that the population of Salzburg Province was 327,232 people—4.719303 percent of the entire population of Austria.

A large distillery once declared that it had squeezed 191,752 oranges, 580,582 lemons, and 453,015 limes to make its whisky sours, daiquiris, and margaritas.

The Federal Government each month provides an enormous quantity of statistical facts about the general economy and major sectors thereof. Many such statistical facts are expressed to an alleged accuracy of one-tenth of a billion dollars. In view of the necessarily crude data gathering, estimation, and compilation procedures employed, the figures are really considerably less accurate than they appear.

This statistical fallacy is found in all branches of statistical investigation. Whatever the context, one is always wise to be skeptical of figures pretending great accuracy. Usually the simple task of asking yourself "Judging from what I know about the thing being measured, can I really believe that such accuracy is possible?" will be sufficient to keep you from being more favorably impressed than you should be.

MEASUREMENT BY PROXY

Sometimes a perfectly good measure is used as in imperfect proxy for something else. The following advertisement, for example, is hypothetical but based upon an actual one that appeared in many magazines:

> Hair Rescue corrects a variety of scalp diseases and stops the hair loss they cause. Hair Rescue has been used by over half-a-million people on our famous Double-Your-Money-Back-Guarantee. Only 1% of these men and women were not helped by Hair Rescue and asked for a refund. This is truly an amazing performance.

The most obviously misleading part of this advertisement is the use of requests for refunds as a proxy measure of the number of customers not helped by the product. What do you suppose is the ratio of dissatisfied customers to the number of customers who actually take advantage of a money-back guarantee? One-to-one as implied by the ad? Two-to-one? Ten-to-one? One hundred-to-one? The true ratio is unknown and quite likely unknowable. Almost certainly, however, the number of dissatisfied customers exceeds the number who demand a refund.

Another satisfied customer.

Who has not heard the oft-repeated advertising claim that nine out of ten doctors recommend the ingredients of a certain patent medicine? It does not necessarily follow that they recommend the product by brand name. I am told by a doctor friend that when the company researchers conduct their surveys they indicate only ingredients, not brand name, and these ingredients are common to many commercial medicines.

The use of proxy measures is often justifiable because the thing of interest just doesn't lend itself to accurate and/or relatively inexpensive direct measurement. However, one should at least be able to recognize when a proxy measure is being used and be prepared to pass judgment on its quality.

Without a doubt, the most insidious proxy measures in the world are never referred to as "proxies" or by any other name suggesting "poor substitute." I am speaking of the proxy information one gets from a *self-selected sample.* When people are asked to write in or call in to express their opinions about, say, a political issue, a court trial, or whether having children is a good idea, the result will be poor proxy information because the opinions of those heard from are almost never representative of the opinions of the much larger group whose opinions are actually sought. We take a closer look at this topic in Chapter 10.

BEWARE OF MIRACULOUS STATISTICAL BIRTHS

Statistical information doesn't simply fall out of the sky into our newspapers or television sets. Somebody has to work to get it—and sometimes the amount of work is substantial. There is always a process involved. Although the specific features of the process may vary from one situation to another, they do have certain things in common:

As noted above, somebody has to define the thing or activity to be measured.

Somebody has to determine where the information resides, i.e. on time cards, in minutes of meetings, in the files of businesses or government agencies, on questionnaires filled out by customers, inside the heads of people, and so on.

Somebody has to determine how to extract the information from its place of residence without jeopardizing its accuracy too much.

Somebody has to decide on alternative ways of extracting the information if the most obvious way is too costly, impractical in some other respect, or might prove deficient from the standpoint of maintaining accuracy.

Somebody—usually a lot of somebodies—have to gather, organize, and summarize the data.

Somebody probably has to put many "little" statistical facts together to get the "big" ones that eventually reach us.

I urge you to begin asking yourself, "What must have been the process through which this statistic made its way from definition to extraction to compilation to 'fact'?" Many statistical "facts" melt and die like the wicked witch in the Wizard of Oz the moment that question is raised. Often the mere act of imagining the prodigious efforts that would be required to get some kinds of statistical facts produces a ludicrous mental picture. Some things are, practically speaking, unknowable; to pretend otherwise is folly.

In Chapter 1, for example, you were asked to evaluate the following statement made by a Kansas City deputy sheriff: "Satanists are responsible for as many as 50,000 deaths a year, mainly transients, runaways, and babies conceived for the purpose of human sacrifice."

Despite this assertion and similar ones blathered out over the airways by legions of TV-tabloid commentators, the idea that human sacrifices are being carried out in connection with satanic rituals may be completely false.

Yes, there have been a few people found guilty of murder who claimed to have been influenced in some manner by satanic beliefs. But it is right there that the hard evidence apparently stops. Evidence of intentional, organized, well-attended, ritualistic murder simply does not exist.

Is that because satanists are extremely clever, as some people maintain? (It's possible, I suppose. In Virginia, in late 1988, a *Style News* article described occult paraphernalia left at a popular riverside park. But a park official wasn't worried; he stated, "The paraphernalia couldn't be the work of real satanists because real satanists don't leave any traces.")

Think about it. The more clever "real satanists" are, the more worthless the 50,000 estimate is. Moreover, anyone willing to search the media will soon find estimates both much higher and much lower than 50,000. We have here, I believe, a "statistic" used for dramatic effect only—not for the dissemination of real information.

But let us suppose, for the sake of argument, that such ritualistic murders do occur in the United States. How do you think someone might go about getting aggregate data on an activity that would be a capital crime for every adult involved? Try thinking it through and visualizing the process, and you will soon find yourself with enough black humor to launch a career in stand-up comedy.

Make up your own fantasy. Just don't forget to address the question: How does the Kansas City Police Department, or any other law enforcement agency for that matter, find out about the number of such crimes—a number used rather freely in police workshops and interviews with the press—without, apparently, ever managing an occasional arrest? The mind reels. Are there undercover police infiltrating the local covens, witnessing such murders, and still coming up short on evidence for arrests? Come on!

Like Athena in Greek mythology, some statistics are born full-grown out of somebody's head. And like Athena, they too are "myths." The task of getting reasonably accurate statistical information about some things is at best difficult and at worst impossible (hence the moniker *unknowable statistic).* On the other hand, making up a "statistic" is a cinch. It is done all the time!

For example, I have on my bookshelf a little paperback that has been sitting there for as long as my green-and-red double-knit suit has been hanging in the closet. It is titled *Criswell Predicts.** Prophet Criswell made a lot of startling predictions and boasted an 87 percent accuracy rate—or so it says in 60-point type on the book's cover. Since quite a lot of time has passed since the book's publication, I recently tried to appraise his accuracy myself: Did an earthquake virtually wipe out San Francisco on April 7, 1975? No. Was Cuba's Fidel Castro assassinated on August 9, 1970? No. Was the earth totally free of rain for ten months in 1977? No. To make a long story short, I have been through the book twice and am still looking for a single accurate prediction. You tell me, if you can, where the "87 percent" came from. Do you suppose it might have been—gulp!—just a lie?

Let us now turn to a consideration of how legitimate statistical facts are sometimes improperly communicated through what I call *cheating charts.*

*Grosset and Dunlap, 1969.

3

CHEATING CHARTS

> The [newspaper] industry has developed recognized standards to measure the quality of writing, headlines, and photographs, but there is little agreement on the standards by which to measure the quality of information graphics. ... We need to know when color and art work contribute to comprehension and when they contribute to confusion.
>
> —Daryl Moen*

It was a pair of charts that first startled me into a realization of how brazenly statistics are sometimes used to deceive the public. It was the same pair of charts that launched me on my St. George-like odyssey in search of statistical dragons to slay—and eventually to the coveted title of Master Statistical Critic.

You see, I was a graduate student seeking part-time work. The prospective job that appealed to me most, until my disillusionment, was that of mutual fund salesman for a company that enjoyed a fine reputation for both integrity and financial success.

*"Misinformation Graphics," *Aldus Magazine* [now *Adobe Magazine*], (Jan/Feb 1990, Volume 1, Number 2), p. 64.

An important visual aid used by this company's salespeople was a film of still pictures designed to show prospective investors the benefits of buying into this mutual fund right away.

Among the pictures was a chart of a rapidly rising line depicting the company's assets per share over several recent years. The line was zooming upward but, unfortunately, the speed of ascent was being given a boost by some tomfoolery with the vertical axis. Cartoons suggestive of the great affluence to be enjoyed by the investor in this fund were placed at intervals along the vertical axis in much the same way as in Figure 3-1 on the following page. (Compare, for example, the space between the numbers "3" and "4" with the space between numbers "4" and "5.")

The cartoons did enhance the entertainment value, and maybe even the aesthetics, of the chart, but they also contributed to the impression that assets per share had been racing upward at warp speed.

In the same filmstrip, another chart was used to make the viewer aware of the menace of creeping inflation. This chart, quite properly, showed the Consumer Price Index, used as a measure of the cost of living, rising over time, as it had been for several recent years. So far, no problem. Unfortunately, the same chart showed a declining line depicting the Purchasing Power Of The Dollar Index. Together, the two indexes presented a picture much like that of Figure 3-2 on page 58. (Figure 3-2 has been made much simpler in construction than the one in the filmstrip so as to illustrate the point in an uncluttered way.)

Actually, the two indexes measure the same thing, the Purchasing Power Of The Dollar Index being essentially the reciprocal of the Consumer Price Index. Only one of these lines is needed to give a proper visual indication of inflation.

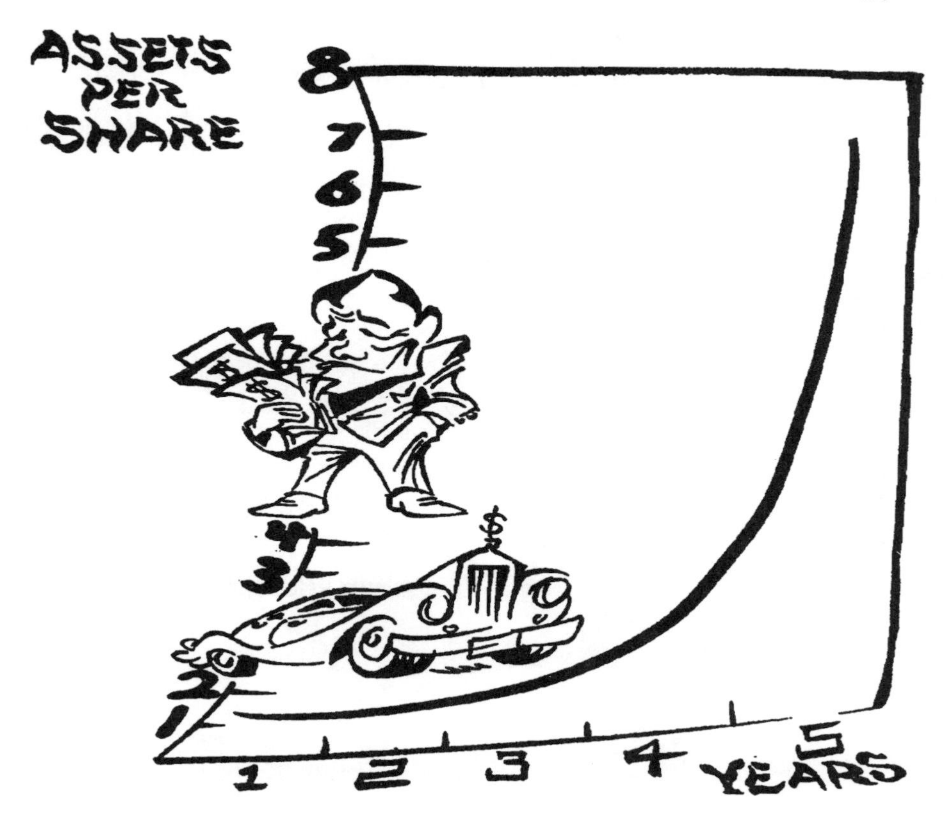

FIGURE 3-1

Use of both lines amounts to intentional—and shamefully misleading—double counting.

It is probably safe to say that no statistical tool has been used more often and with greater success to deceive the unwary than the statistical chart. And don't be too quick to conclude that intentional deceit is the only trap in the path of one innocently seeking enlightenment from such a chart. Charts represent someone's attempt to communicate statistical information in a more comprehensible and more interesting way than would be possible with the raw data alone. Even though that intention is certainly praiseworthy, the harsh reality is that such attempts at clarification often backfire. As the quotation introducing this chapter suggests, confusion is often found where clarification was supposed to reign; what the creator of the chart intended and what the viewer actually internalizes are sometimes two very different things.

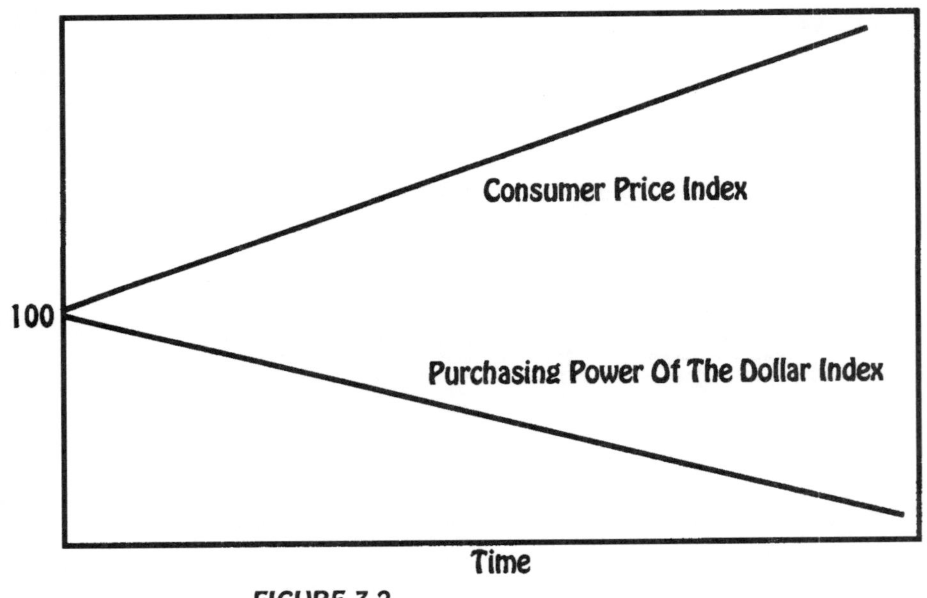

FIGURE 3-2

To make matters still worse, sometimes our own vision will play tricks on us. Please humor me momentarily while I help you recall some classical optical illusions which you certainly must have mused over, as I did, during childhood: Figure 3-3, for example, shows a rectangle whose border lines are obviously straight. However, the two center horizontal lines appear curved. A check with a ruler will quickly assure you that they too are straight.

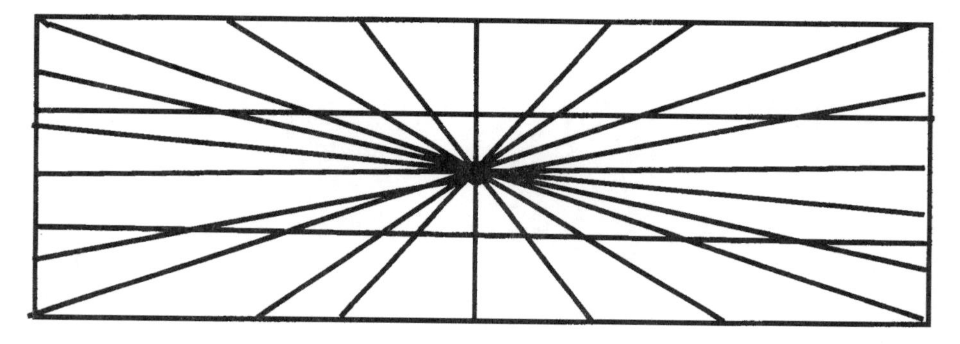

FIGURE 3-3

In Figure 3-4 the "egg-shaped" things in the center are perfect circles, though they certainly don't look it.

FIGURE 3-4

Figure 3-5 shows a tiny model of St. Louis, Missouri's famous Gateway Arch. This arch appears to be much taller than it is wide, even though its height and base are equal.

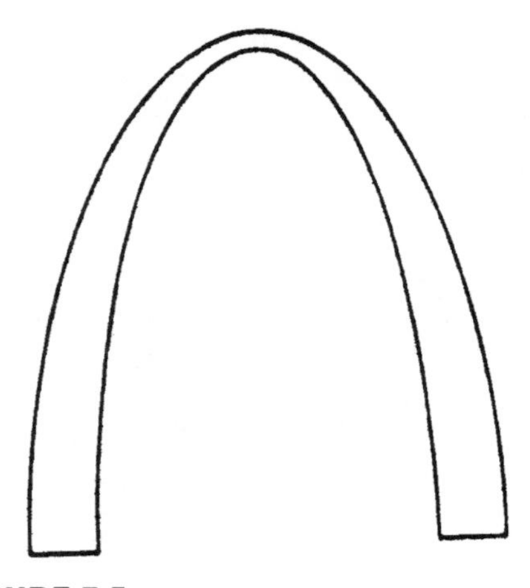

FIGURE 3-5

One of the really strange things about such optical illusions is that, even when we know the truth, we cannot make a mental adjustment that will render the picture "correct." It looks the way it does and that's that. But do such optical illusions have anything to do with statistics? Indeed they do—or, at least, they can if we aren't very careful. For example, examine Figures 3-6 through 3-9.

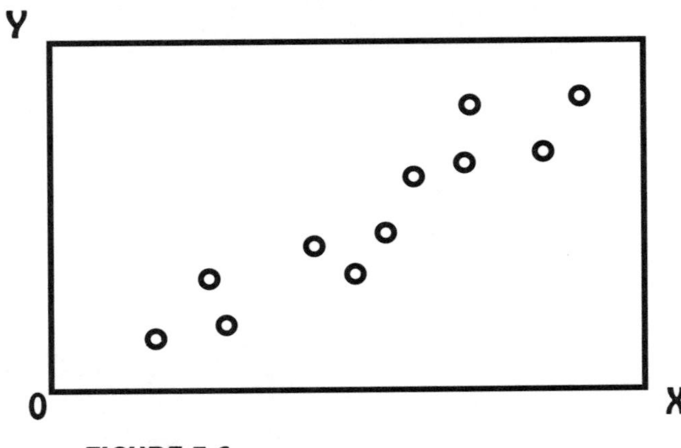

FIGURE 3-6

Figure 3-6 illustrates a most useful graphic tool which statisticians call a *scatter-diagram.* (More is said on this subject in Chapter 11.) A scatter-diagram can give the viewer a quick indication of how strongly two variables are related. For example, let us call one variable Y and measure it on the vertical axis. Let us call the other variable X and measure it on the horizontal axis. Now, if we indicate *pairs* of Y and X values with little circles, a graph like that in Figure 3-6 emerges.

Figure 3-6 shows what is probably a rather strong relationship, an impression gleaned by noting how close the various little circles are to an imaginary line running through their center. To help you understand this idea, I had my loyal computer fit a mathematical line to the data on which Figure 3-6 is based; the result is shown in Figure 3-7. See how close to the line the circles tend to lie. The more closely the circles hug the line, the stronger usually is the relationship. Naturally, a strong relationship is generally preferred over a weak one.

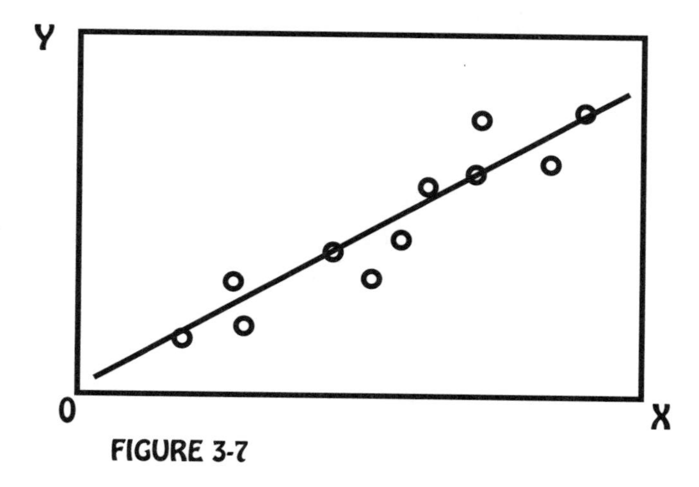

FIGURE 3-7

You now know how to use a scatter-diagram to distinguish between a strong and a weak relationship. Right? Not so fast! Optical illusions, it seems, can infiltrate even the most sacred realms of statistical graphics:

A research study was conducted using scatter-diagrams similar to those shown in Figures 3-8 and 3-9.* The one in Figure 3-8 is constructed so that the collection of little circles occupies only a small part of the space available. Figure 3-9 shows *exactly the same* data, only this time, the small circles nearly extend over the entire space.

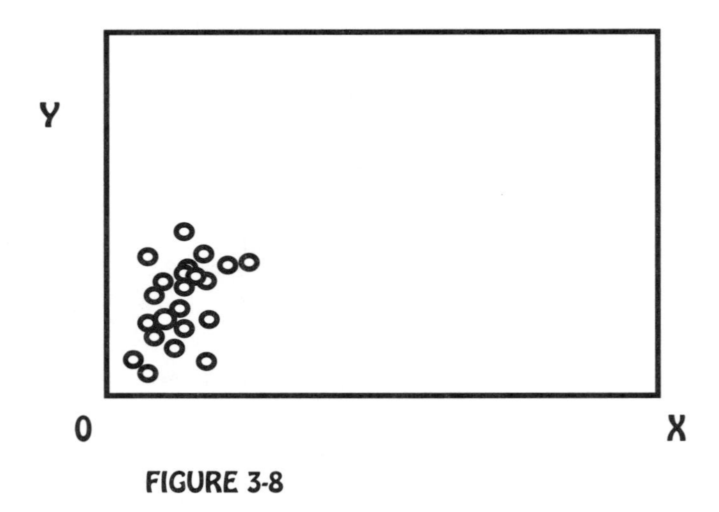

FIGURE 3-8

Professional statisticians and other scientists with statistics training were asked to judge just how strongly the two variables are related. Most judged a small plot within a large field (as in Figure 3-8) as indicating greater strength than a large plot within an only slightly larger field (Figure

*See Cleveland, W. S., Diaconis, P., and McGill, R. (1982). *Science,* **216,** 1138-1141.

3-9). For relationships of middling strength, the manner of plotting the data altered perceptions of strength by as much as 15 percent.

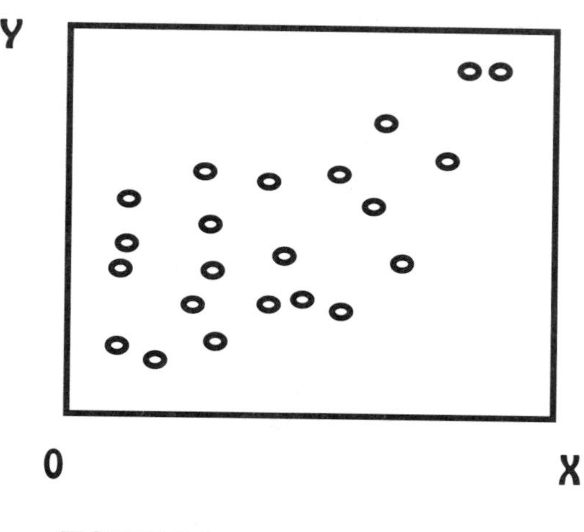

FIGURE 3-9

This example, like many yet to come, shows that it sometimes takes serious thought and effort to keep visual gremlins out of statistical charts. Unfortunately, serious thought and effort are now the very things most rapidly disappearing from chart construction.

If anything, the present chapter has gained in relative importance since publication of my earlier book, *Flaws and Fallacies in Statistical Thinking,* a quarter of a century ago. Today, with personal computers and laser printers conveniently available at affordable prices and with new software making it possible for anyone with the motivation to produce beautiful two- and three-dimensional graphics or dazzling slides in practically no time at all, we are up to our ears in breathtaking charts. But remember: Modern technology cannot make a chart honest if its creator wants it dishonest. Nor can modern technology indemnify a chart against unintentional communications glitches or optical illusions. It is perhaps safe to say that such errors are more the rule today than they were back in the days when charts had to be constructed with a more "hands-on" approach.

Clearly, where charts are concerned, the statistical critic faces a multitude of challenges. Let us look at some more examples.

THE EYE-POPPING LINE CHART

Table 3-1 on the following page contains a fairly large set of numbers. These particular numbers represent quarterly values of Standard and Poor's Index of 500 Common Stock Prices for the 14 years from 1981 through 1994. What do you make of them? Unless you already have a trained statistical eye, they probably don't tell you anything except, perhaps, that the general drift of stock prices has been upward over the years.

The difficulty of making sense out of a mass of data like this is precisely why charts are so often used. Tables overflowing with figures simply serve up too much detail for the mind to assimilate.

Figure 3-10 on page 66 shows the same data in the form of a line chart. In this chart not only is the long-run, upward trend of stock prices clearly revealed, but so also are some shorter-term swings and even-shorter-term zigs and zags. One sees at a glance a pretty thorough display of what has happened to stock prices over this period of time.

This figure utilizes an arithmetic scale on the vertical axis. With such a scale, equal numerical changes in the data show up as equal-sized vertical movements. Often data shown over

Table 3-1. Standard and Poor's Index of 500 Common Stocks, by Quarter, 1981-1994

Year And Qtr.	Index	Year And Qtr.	Index	Year And Qtr.	Index	Year And Qtr.	Index
1981-1	131.5	1984-3	160.5	1988-1	258.1	1991-3	385.6
1981-2	132.8	1984-4	165.2	1988-2	263.1	1991-4	387.1
1981-3	125.7	1985-1	177.3	1988-3	266.9	1992-1	412.0
1981-4	122.2	1985-2	184.8	1988-4	275.0	1992-2	410.2
1982-1	114.2	1985-3	188.3	1989-1	290.7	1992-3	417.2
1982-2	114.1	1985-4	197.0	1989-2	313.3	1992-4	423.7
1982-3	113.8	1986-1	220.0	1989-3	342.0	1993-1	442.4
1982-4	136.7	1986-2	240.6	1989-4	345.4	1993-2	445.5
1983-1	147.7	1986-3	241.2	1990-1	336.3	1993-3	453.6
1983-2	162.7	1986-4	243.7	1990-2	349.6	1993-4	464.2
1983-3	165.5	1987-1	279.3	1990-3	335.4	1994-1	469.5
1983-4	165.7	1987-2	293.3	1990-4	317.1	1994-2	451.0
1984-1	160.4	1987-3	319.4	1991-1	353.3	1994-3	460.9
1984-2	155.8	1987-4	255.4	1991-2	378.7	1994-4	460.0

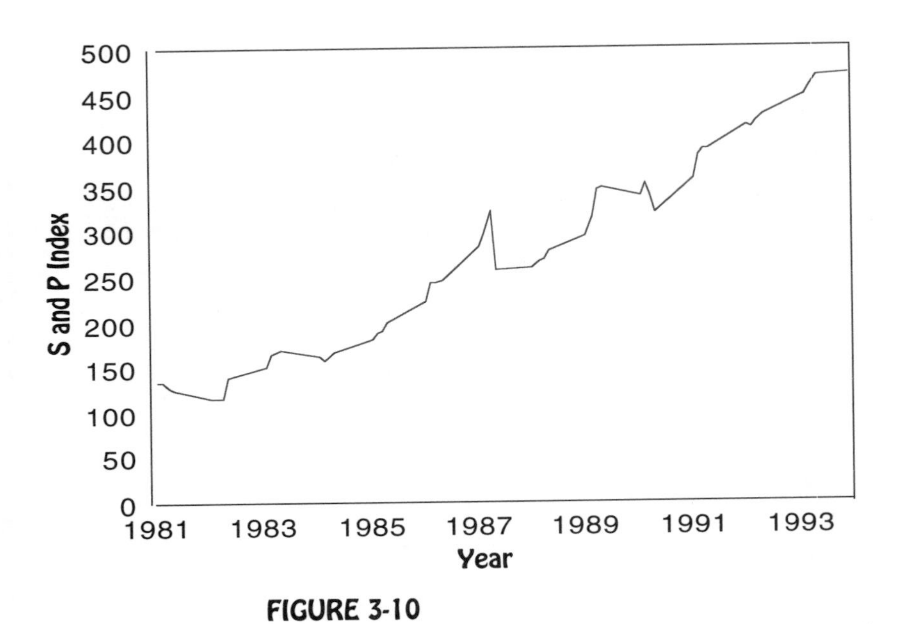

FIGURE 3-10

relatively long stretches of time are charted in a different manner: The vertical axis will be manipulated (legitimately) so that equal *percent changes* in the data, rather than equal numerical changes, are seen on the chart as equal-sized vertical movements.

Figure 3-11 on the following page shows these same numbers charted in that way. The vertical axis has a logarithmic, or ratio, scale. That is, the vertical-scale values are spaced in accordance with differences in their logarithms, rather than in accordance with their actual numerical differences. The horizontal axis has a regular arithmetic scale. This combination of a logarithmic scale and an arithmetic scale makes Figure 3-11 a *semilog chart.*

If you don't happen to feel at home with logarithms, it matters very little. More important to the statistical critic is the ability to distinguish between a totally arithmetic chart, on the one hand, and a semilog chart on the other. An examination of Figures 3-10 and 3-11 should help. Especially note the following two features of Figure 3-11: (1) The lines corresponding to 200, 300, and so forth, are associated with progressively smaller vertical distances as we scan the chart from bottom to top and (2) The vertical axis begins with a

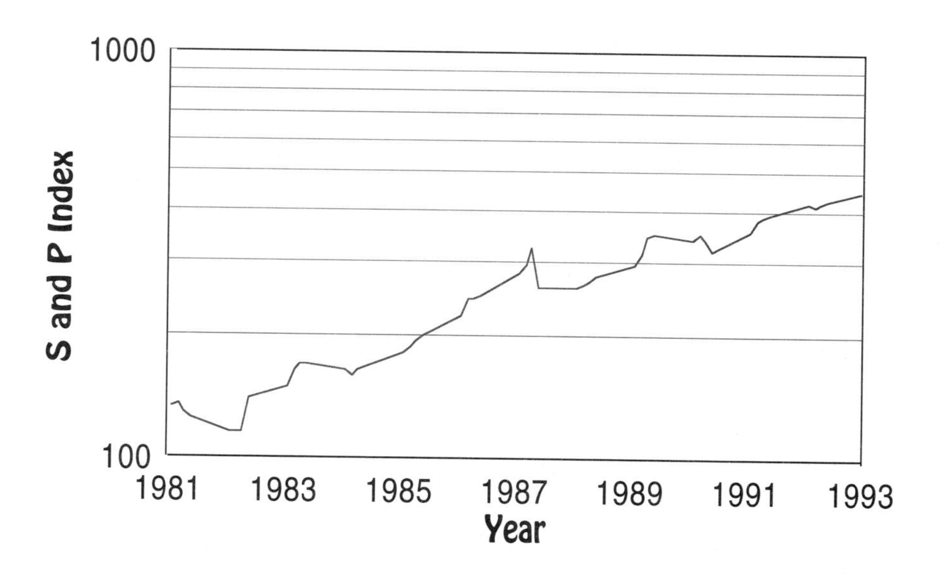

FIGURE 3-11

positive, nonzero number—100 in this case. Log scales cannot begin with zero.

Use of either kind of scale, arithmetic or logarithmic, on the vertical axis, is perfectly legitimate. The choice will depend on whether one wishes to present a picture of numerical or relative changes.

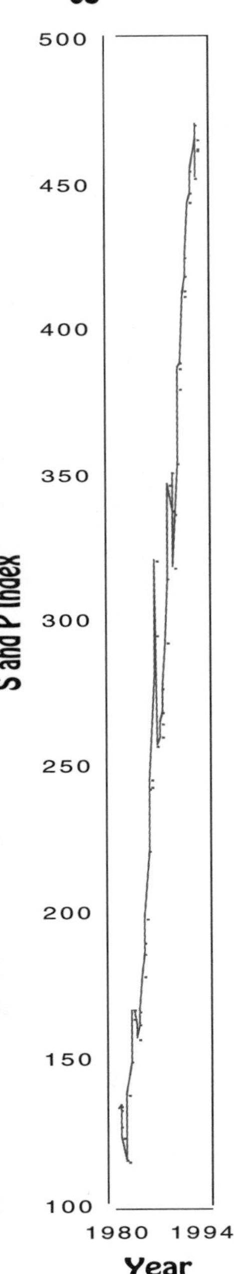

But watch out—

With either kind of vertical scale, in the few seconds it takes to change a couple of computer commands, we can alter the appearance of the stock index data in just about any way we like: For example, if we had a vested interest in convincing someone that stock prices had risen extremely vigorously over the 14 year period all we would have to do is stretch out the vertical axis like crazy (and *truncate* it, that is, cut a little off the bottom) and compress the horizontal axis. This has been done in Figure 3-12.

FIGURE 3-12

On the other hand, if we wished to play down the growth in stock prices, we could do our stretching and compressing in the opposite way, as in Figure 3-13.

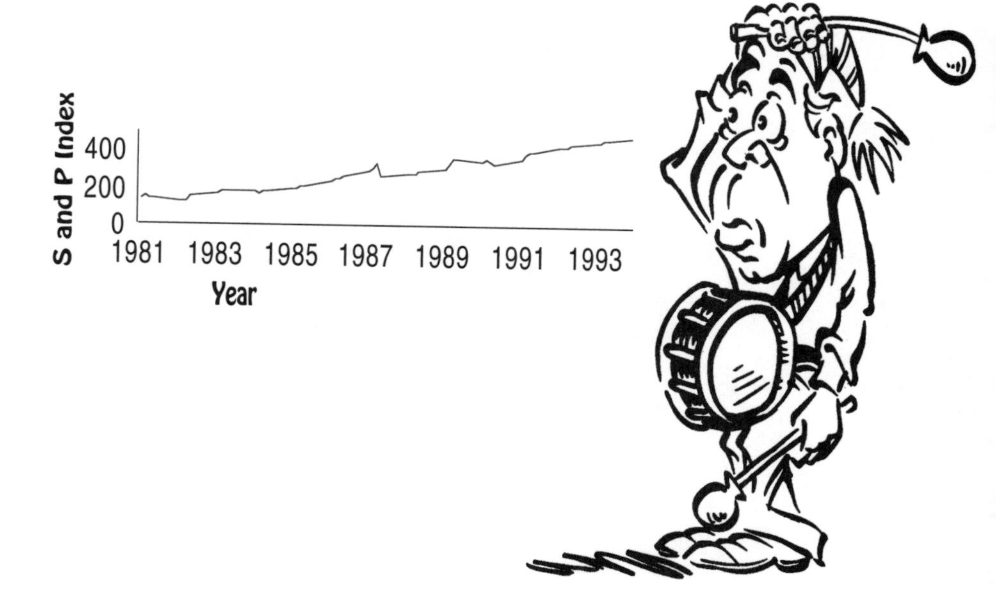

FIGURE 3-13

Actually, if we were dead serious about making the growth in stock prices look as anemic as possible, there are two additional tricks, illustrated in Figure 3-14, that we shouldn't overlook: (1) Flatten the growth path a little more by using a log-scale Y axis and (2) start with the year 1987 rather than 1981. (No one ever said the beginning year had to be 1981 or any other specific year; beginning with 1987 makes it look as if there had been virtually no growth at all during recent times.) Now, take another look at Figure 3-12 and compare it with Figure 3-14. Isn't it amazing how different the same set of data can be made to look? And, gee whiz, nowadays we don't even have to do our own lying; the computer will do it for us and suffer none of the old irrational hang-ups related to conscience and stuff like that.

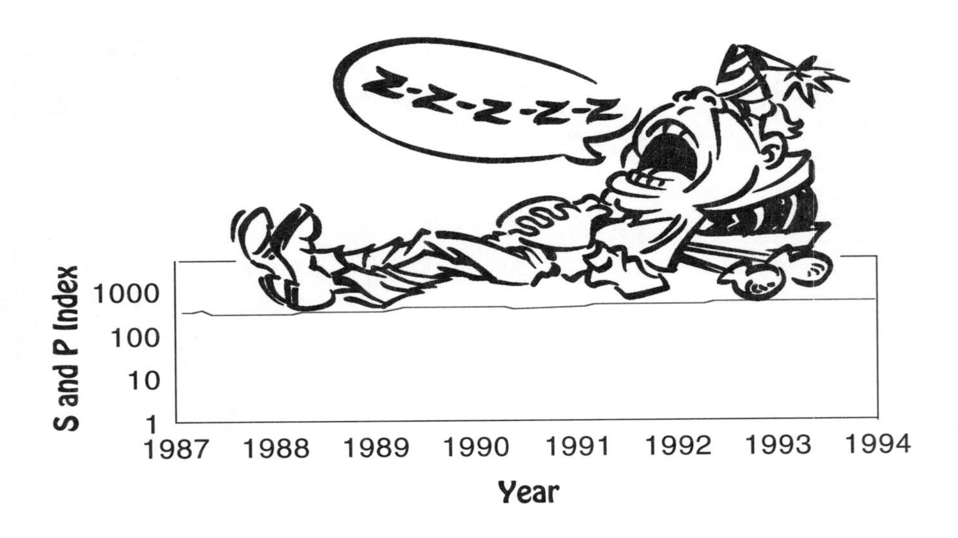

FIGURE 3-14

MORE ON LOG CHARTS

We passed rather hurriedly over the advantages that sometimes result from the use of a logarithmic, rather than an arithmetic, scale on the vertical axis. Some further elaboration seems in order.

Suppose that *Better Houses* magazine has been preeminent in its field for a great many years. Suppose further that five years ago, *Homes and More,* a direct competitor, appeared on the scene. Finally, suppose the recent five-year history of numbers of subscribers to the two magazines has been as shown in Table 3-2.

Table 3-2. Number of Subscribers to Two Magazines For A five-Year Period (Thousands)

Year	Better Houses	Homes & More
1	8932	0.2
2	9111	0.5
3	9293	1.2
4	9479	3.0
5	9668	7.0

If subscription figures were to be used in advertisements for the two magazines, *Better Houses* would benefit from showing the data with an arithmetic vertical scale. Figure 3-15 conveys the definite impression that *Better Houses* is doing better than its younger competitor, not only in terms of absolute number of subscribers, but in terms of year-to-year increases as well.

Keep in mind, however, that *Homes and More* is only five years old. No one could reasonably expect it to be performing on a par with its older competitor. Interest, therefore, should center neither on absolute level nor absolute increases in subscribers. Relative increases are probably the most meaningful data to compare. If *Homes and More* were to use these subscription figures in its advertising, it would be well-advised to use a logarithmic

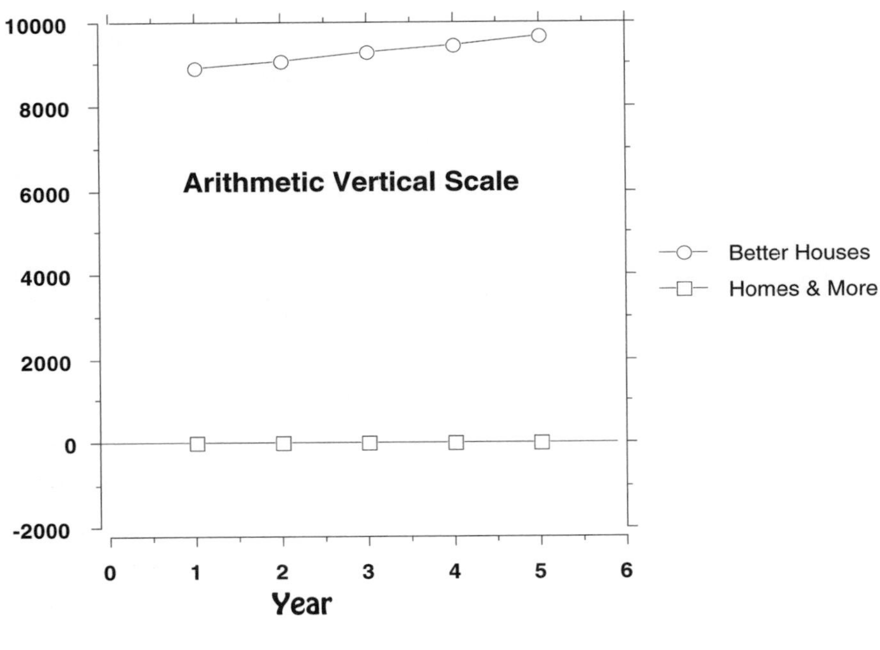

FIGURE 3-15

Y-axis to show that its relative increases have been greater than those of *Better Houses*. The faster relative growth of the younger magazine shows up dramatically when plotted in that manner, as Figure 3-16 demonstrates. When viewed in this light, *Homes and More* is seen to be encroaching rapidly on the older magazine's turf.

None of the preceding is meant to suggest that arithmetic charts are inherently bad or that semilog charts are inherently good. To repeat a point made earlier: the choice will depend entirely upon what one wishes to convey with the chart. In this example, relative increases were assumed to be of paramount importance; therefore, a semilog chart was the more appropriate choice.

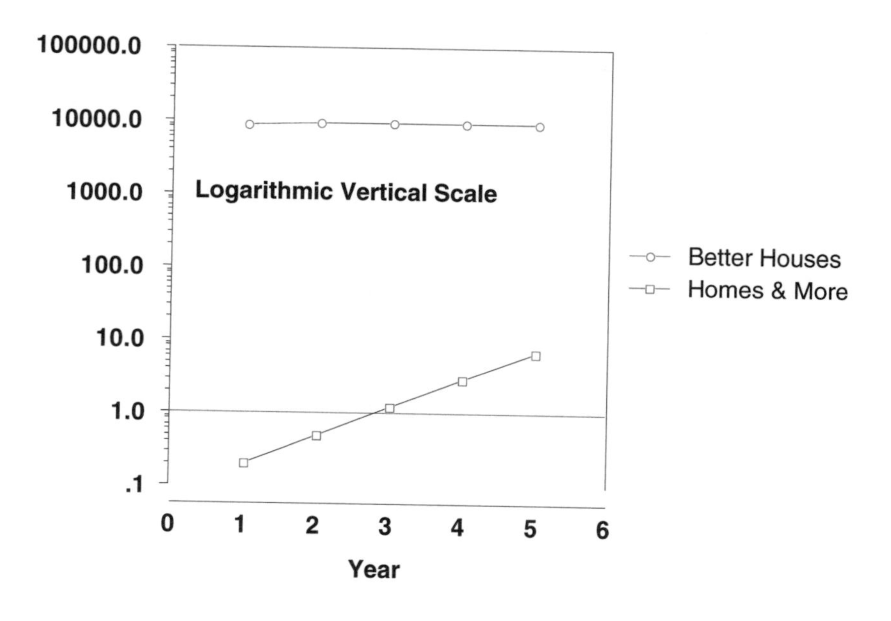

FIGURE 3-16

Unfortunately, the mere fact that creators of statistical charts have a choice of scale types opens the door for possible abuse. As a statistical critic, you will be wise, when confronted by a chart prepared by someone else, to ask yourself whether the impression conveyed would be materially different if the other kind of scale had been used.

UNMARKED AXES

In view of what has been said about line charts, it almost goes without saying that charts with unmarked axes should never be trusted. When the creator of a chart is free of the discipline imposed by well-marked axes, he or she can convey through the chart just about any impression desired. In Figure 3-17, for example, we see two sales lines rising at different rates. Or do we?

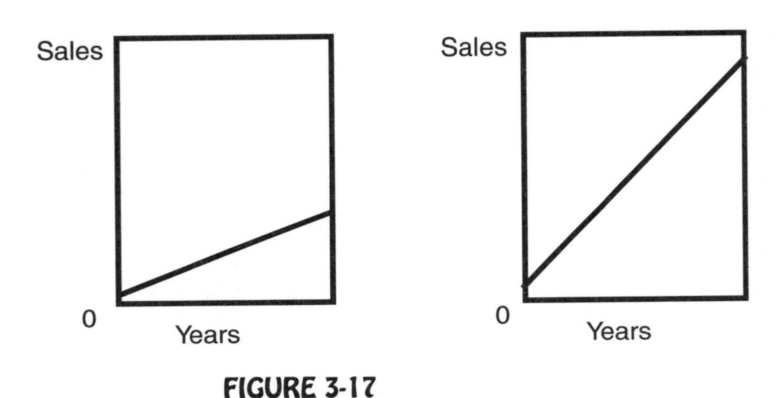

FIGURE 3-17

When numbers are placed along the vertical axis, as in Figure 3-18, they reveal that the two sales lines have advanced at close to the same rate. The one on the right simply has a more stretched-out vertical scale.

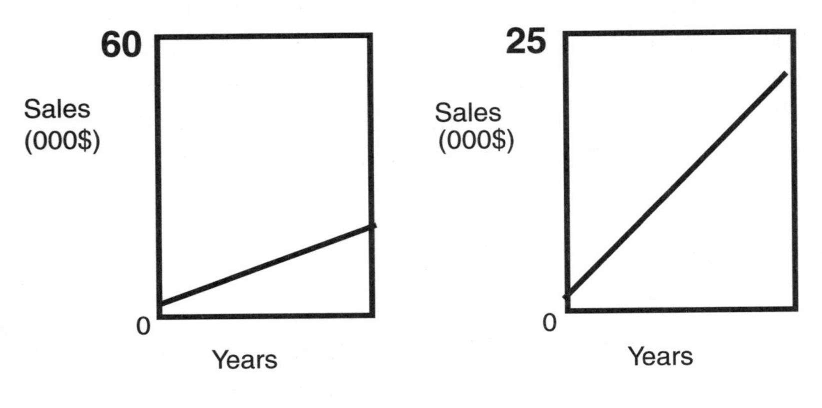

FIGURE 3-18

CHARTS THAT WON'T BE RIGHT WHATEVER YOU DO

As this is being written, much public concern is being expressed over the soaring costs of health care. In this section we will use some actual health care data from *American Demographics* Magazine to highlight the problem and to demonstrate columns charts and pictograms.

Between 1965 and 1987, per capita consumption of personal health care for people aged 65 and over virtually tripled, rising from approximately $1800 to about $5400. (These figures are expressed in terms of constant 1987 dollars.) These are the numbers which will be used for purposes of illustration. The two expenditures values are shown in standard columns-chart form in Figure 3-19.

Just as long as the widths of the bars are equal, as they are in Figure 3-19, this columns chart allows us to compare health-care expenditures for the two years simply by eyeballing the heights of the bars. Of course, the bottoms of the bars shouldn't be chopped off, as they often are, or the visual comparison becomes invalid.

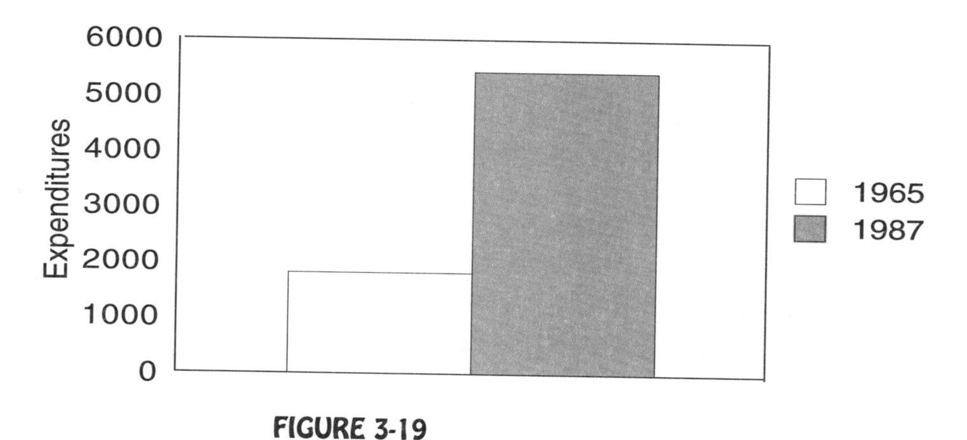

FIGURE 3-19

The only trouble with columns charts, or so many people seem to believe, is that they are terribly uninteresting things to look at. As a result, creators of charts often use pictures of related objects to tell their stories. Use of such *pictograms,* as these more entertaining charts are called, does indeed enhance the eye appeal of the presentation; but it also usually introduces some unavoidable technical problems. For example, if the amount spent on personal health care in 1965 were indicated by a prescription bottle like that shown in Figure 3-20, how should the amount for 1987 be shown?

FIGURE 3-20

Should it be shown using three prescription bottles of the same size, as in Figure 3-21?

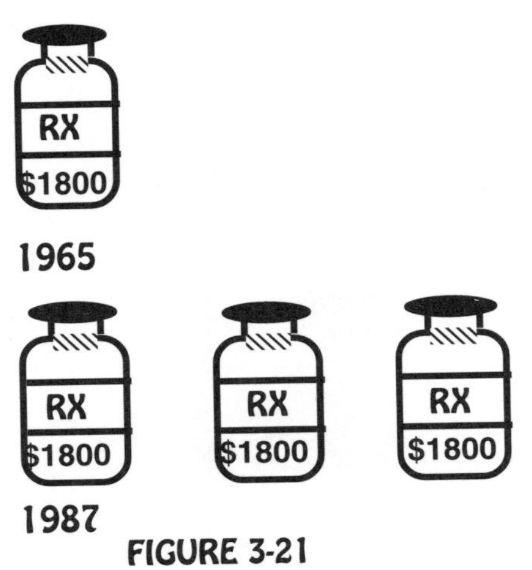

FIGURE 3-21

Such a presentation has at least two shortcomings. First, the 1987 value is not shown directly and must be calculated by the viewer. Second, the viewer might misinterpret the message as :"Whereas the typical senior citizen took only one kind of prescription drug in 1965, he or she was taking three kinds of prescription drugs in 1987." This is, admittedly, something of a stretch, but a conclusion easily reached by anyone not carefully scrutinizing the chart and the accompanying explanatory material.

An alternative way of presenting the data, one more in accord with the columns-chart approach, is to show the 1987 prescription bottle three times as tall as the 1965 bottle. But this procedure presents problems of its own, as can be seen in Figure 3-22.

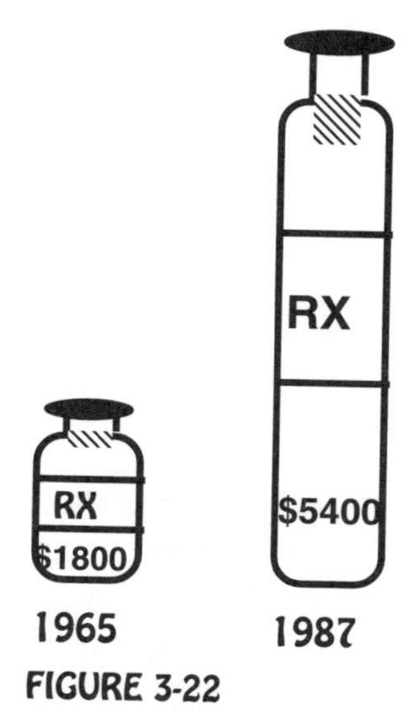

FIGURE 3-22

Figure 3-22 is accurate in its way, but the 1987 bottle looks pretty strange. Why go to the trouble of making a picture just to have it look strange? Of course, one can easily get around this kind of strangeness by making the second bottle three times as wide (in addition to three times as tall) as the 1965 bottle, as shown in Figure 3-23. Now the second one looks more like the bottles we're used to seeing, but a serious technical error has been introduced. Considering only the two dimensions, height and width, the second bottle is seen to be nine times, rather than only three times, as large as the first!

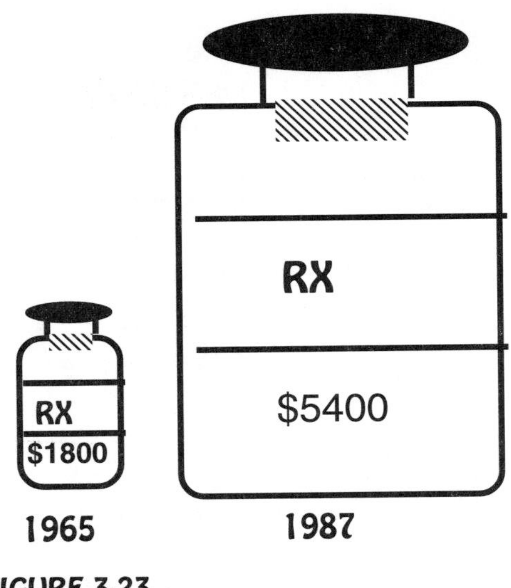

FIGURE 3-23

To change the chart so that the areas are correct in two dimensions, the 1987 bottle should presumably be the square root of 3 = 1.73 times as high and 1.73 times as wide as the 1965 bottle, as in Figure 3-24. But that still isn't the whole story. Because the picture is really of three-dimensional objects, the viewer who looks at it in that light, can easily get the impression that the total volume represented by the second bottle in Figure 3-24 is far greater than it should be.

Thus, three dimensional accuracy would require our making both the height and width of the second bottle still smaller, relative to those of its counterpart. How much is "still smaller?" Well, the formula for the volume of a cylinder could help us find out—if we were really that concerned about knowing. Regrettably, further fussing with the height and width will not necessarily result in a better chart. One simply never knows whether the viewer will interpret the picture as three

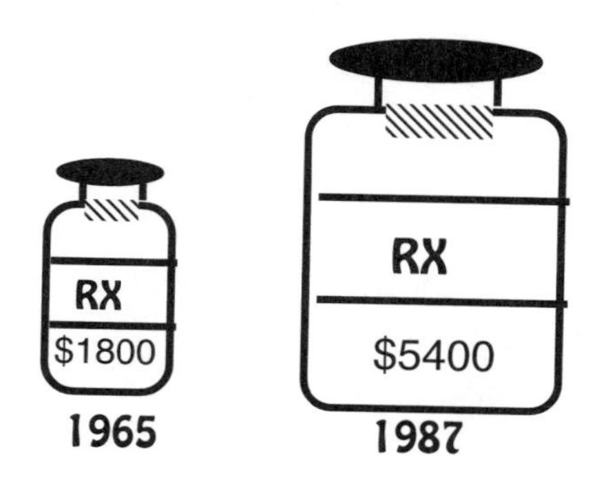

FIGURE 3-24

dimensional because the object is three dimensional in real life, or as two dimensional, because that is how many dimensions it has on two-dimensional paper.

There is no doubt about it: Pictures do dress up a chart. Other things being equal, I personally prefer a dressed-up chart to a more austere one. Alas, in chart construction other things are seldom equal, and a little cuteness can exact a heavy toll in accuracy. The chart maker aspiring to create a *visually accurate* pictogram has about as much chance of succeeding as he has of winning the Powerball Lottery—twice!

Call me an enemy of the arts if you like, but I would still advise you to leave measurement-related pictures off your own charts. (Cute stuff having no measurement function is okay.) When you run across pictograms prepared by others, you should examine them with care. Maybe someone is trying to help you get a wrong impression. Even if that is not the case, you can easily mislead yourself by interpreting the pictogram in a manner different from that intended.

Much more could be said about charts, of course. But we have considered some of the most common pertinent hazards. Points made in connection with charts presented here could easily apply to other kinds of charts as well. Let us, then, put the subject to rest and proceed to a consideration of descriptive statistical measures. The following three chapters will introduce you to averages, measures of variation, and percents in that order.

4

ACCOMMODATING AVERAGES

> When I was a young man practicing at the bar, I lost a great many cases I should have won. As I got along, I won a great many cases I ought to have lost; so on the whole justice was done.
>
> —Lord Justice Matthews

The story is told about a traveling man who registered at a posh hotel. A bellhop carried the man's luggage to his room and waited for a tip. Catching the hint, the man asked, "How large a tip do guests of this hotel usually give you?" "Ten dollars is about average," the boy replied without hesitation. The man, obviously surprised, removed a ten-dollar bill from his wallet and handed it grudgingly to the bellhop. The boy was delighted and informed the man, "You're the first one ever to come up to my average."

The word *average* originated as a clearly defined insurance term based on the fact that many kinds of data tend to cluster around some central, or representative, value. But today, as the bellhop story suggests, the term has so many diverse meanings that a number purported to be an average can be very confusing in the absence of further clarifying details.

PHANTOM AVERAGES

Let us dispose of two misuses of the word "average" right away. First, to speak of an "average man" or an "average woman" or "average weather" or an "average movie" or an "average whatever," is to speak nonsense, even though it is commonly done.

A person might be average with respect to height, weight, income, intelligence, strength, or some other identifiable, and measurable, characteristic. However, the term "average man" conveys nothing except perhaps the vague notion that the man in question is not outstanding in any way that meets the eye. Similarly, the weather might be average with respect to temperature, rainfall, or humidity, but not just "average."

Second, even when one works with a specific quantifiable property, it is not inevitable that the resulting "average" will convey any real information. For example, suppose you know a doctor who earns $300,000 a year. This doctor has a wife without an independent source of income. He also has four children, three of whom earn no money. The fourth child earns about $2500 a year mowing lawns. The doctor hires a part time housekeeper and a part time gardener and pays them $5000 and $6000 a year, respectively.

Granted, there are numbers here that can be averaged—and they all even pertain to income. Just sum the eight numbers and divide by eight: ($300,000 + $0 + $0 + $0 +$0 + $2500 + $5000 + $6000)/8 = $39,187.50. Yes, the average income for these eight people is a respectable $39,187.50—but what does it mean?

Half the people involved really earn nothing at all; should they be part of an average income calculation? Maybe so. But the fact is not obvious. Two of the people whose incomes are reflected in the average aren't even members of the family.

Indeed, their incomes are really "outgo" as far as this family is concerned. Do they deserve to be represented in the average? Again, it is not obvious that they do. What we have here is an example of what I call a *pseudoaverage*, an average determined from a collection of numbers too disparate in what they represent to permit a result clearly expressible in words.

In this chapter, we will be limiting our attention to true averages, as distinguished from these pseudoaverages. Nevertheless, the term "average" can still be very confusing because there are many kinds of averages that can be employed in any given situation. From a careful statistical standpoint, some averages are better suited to certain data sets than others. By the same token, some are better suited to presenting a distorted picture of the data than others.

Clearly, one should definitely not just jump into the calculation of an average—or allow someone else to do so. The kind of average selected should represent the total body of data in the least distorted way.

The desirability of selecting an appropriate type of average before beginning calculations may seem self-evident. Nevertheless, it is a consideration ignored by many people who simply assume that an average is an average and "Why make a big thing out of it?"

The other side of the story is that there are many people who are keenly aware of the several averages in the statistician's tool kit but who lack the statistician's ethics. This is the group that intentionally selects the most misleading average to present the data in a corrupt way. We must be wary of the statistical products of both groups—the uninformed and the unscrupulous.

CHOOSING AN AVERAGE FOR MAXIMUM IMPACT

Let us assume that the latest issue of *Skylarker* magazine contains an eye-filling full-page advertisement dedicated to enticing consumer-goods businesses to advertise in the magazine. Assume further that the *Skylarker* ad reads: "Last year, subscribers spent an average of $1200 on gifts." The $1200 may seem impressive, but how much information does it actually convey? It is possible that it conveys no information at all—or even fraudulent information.

I must ask you to accept, for the moment, the preposterous assumption that the magazine has only seven subscribers: Adams, Baker, Clark, Davis, Elsea, Ford, and Gould. Ignoring Ford and Gould for now, let us suppose that the other five subscribers spent the following amounts on gifts last year: Adams, $144; Baker, $144; Clark, $150; Davis, $180; and Elsea, $210.

When one speaks of "average" without qualifying the term, he or she is usually assumed to be referring to what statisticians call the arithmetic mean—the sum of the values divided by the number of values being averaged. In this case, the arithmetic mean is ($144 + $144 + $150 + $180 + $210)/5 = $165.60, a number considerably under the claimed $1200.

Think of the arithmetic mean as the "center of gravity" of the data, as illustrated in Figure 4-1 on the following page. The seesaw is balanced just as long as the tip of the triangle, that is, the arithmetic mean, is at $165.60. If a value larger than $210, say, were introduced, this balance would be upset unless the tip of the triangle were moved to the right to the position corresponding with the new arithmetic mean.

As previously mentioned, several kinds of averages can be obtained from the same figures, two frequently used alternatives being the median and the mode.

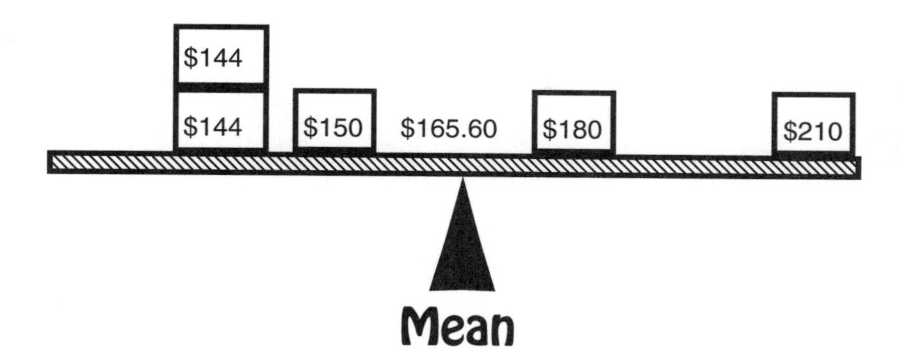

FIGURE 4-1

The median is simply the figure such that exactly half of the numbers are lower in value and exactly half are higher. Applying this definition to our five gift-spending figures, we find that the median amount spent for gifts is $150 because $150 exceeds the two $144's but is exceeded by the $180 and the $210. (See Figure 4-2 below.)

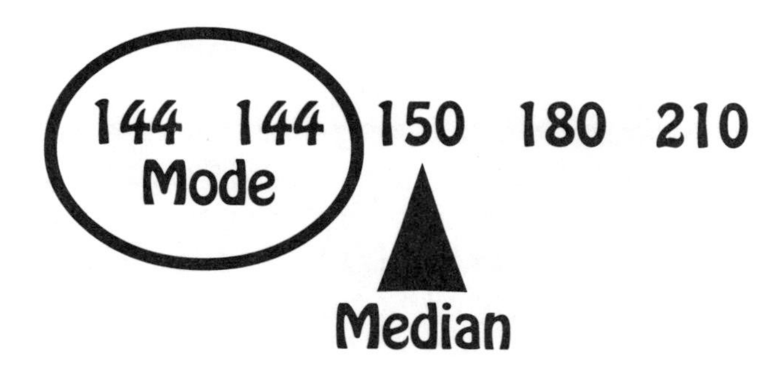

FIGURE 4-2

Finally, as shown in Figure 4-2, preceding page, the mode is the value that occurs most often, namely the $144 in the present case. Thus, we have: mean = $165.60, median = $150, and mode = $144. But which is the most appropriate representative value? The answer depends entirely upon how the measure-elect is going to be used. In this example it probably matters little which average is chosen because the three values are very close to one another. But, of course, all are well below the $1200 claimed in the magazine ad.

But now let's get Jane Ford, a rather affluent manager, into the calculations. Suppose Jane spent $900 on gifts last year. We now have six figures—$144, $144, $150, $180, $210, and $900—to be averaged. With the introduction of Jane, the arithmetic mean becomes a fairly impressive $288, exactly double the mode (still only $144) and considerably higher than the median, $165 (the two competing values $150 and $180, having been summed and divided by 2, as is the conventional procedure when working with an even number of values.)

It now matters a good deal which average is used. The arithmetic mean of $288 is higher than five of the six numbers and lower than only one, a condition suggesting that its use as a representative measure could be quite misleading.

Pushing the example to an even more outlandish extreme, let us now bring in Gil Gould, who spent $6,672 on gifts last year. We find that the arithmetic mean jumps to $1200, the figure appearing in the ad, whereas the median is a mere $180 and the mode is still only $144, the same as always!

Whether the median or the mode would be the preferred representative value is a matter for debate (though in many cases the median will be favored simply because the basic data do not always serve up a clear, single number for the mode). One thing is self evident, however; the arithmetic mean is too greatly influenced by the amount spent by Gil Gould to be representative of all seven numbers. Clearly, the arithmetic mean can be very sensitive to the presence of extremes. Very sensitive.

Clearly, the arithmetic mean can be very sensitive to the presence of extremes.

Very sensitive.

We can now relax the assumption about there being only seven subscribers without negating the point just made. Whether the subscribers are few or many, the related gift-giving data will not necessarily be such that mean, median, and mode are approximately equal—as the writer of the ad had perhaps hoped we would imagine. But because expenditures for gifts are tied to income levels, and because an awfully lot of people have lower incomes while relatively few have higher incomes, the arithmetic mean in this case is almost certainly inflated by the affluent minority.

You might be interested to know that, although I made up the numbers used in this example, a very similar, unquestionably misleading, ad ran in a prestigious national magazine for many years.

Let us consider a situation where an average can be misleading in quite a different way. The arithmetic mean was originally intended as a measure of central tendency. Where no central tendency exists, not only the mean but the median and mode as well might be of questionable worth.

For example, suppose realistically, that a certain major U.S. city has, over the course of a year, a distribution of percent cloudiness like the following: Zero- and 100-percent cloudiness occur more often than either 10- or 90-percent cloudiness. These, in turn, occur more often than either 20- or 80-percent cloudiness, and so it goes, down to 50-percent cloudiness, which occurs rather infrequently.

The arithmetic mean and the median of approximately 50 percent could give the false impression that, typically, the sky over this city is about half covered with clouds when in fact it is more likely to be completely covered with them or completely free of them. Probably the most meaningful way to summarize such a situation is to say: "The data are bimodal with modes at 0 and 100."

A good rule of thumb for those occasions when you are told neither the kind of average used nor the appearance of the basic data is to attach no importance whatever to the proffered "average." Someone just might be using a carefully selected "average" as an instrument for exaggerating a point or in some other way conveying a false impression.

In the above—admittedly extreme—examples, I have probably maligned the arithmetic mean more than I should have. The truth is that the arithmetic mean is indispensable to a great many applications of formal statistical analysis. Often, within the realm of descriptive statistics, the arithmetic mean serves nicely as a measure of central tendency. When a measure of central tendency is selected within the realm of inductive statistics, where sample data are used to help investigators draw conclusions about a much larger body of data, the arithmetic mean is the hands-down choice. Explaining why would take us far afield. But look up the *central limit theorem* in practically any basic statistics textbook.

A mean, a median, or a mode can, if judiciously selected, serve as a useful, one-figure summary of a set of data. However, it usually must be supplemented by other kinds of descriptive measures as well. Otherwise, it can conceal as much as it reveals. An average, for example, tells nothing about the amount of variation in the data, a shortcoming whose possible consequences will be explored in the next chapter.

5

AIR, GRAVITY, AND VARIATION

> Two statisticians were drafted into the army and soon found themselves fighting side by side in enemy territory. Suddenly, they both spotted an enemy sniper and fired at him simultaneously. One statistician fired a yard too far to the right and the other a yard too far to the left. Being statisticians, they shook hands and congratulated each other.
>
> —Old story of unknown origin

Some things in life are so ever-present that they almost escape notice. High on my list of such things would be air, gravity, and *variation*. What's that? Variation? Yes, *variation.* It has to do with how greatly like things differ from one another with respect to a specified, measurable property, such as length, cost, intelligence, mileage, or something else.

Variation is, without a doubt, one of the most ubiquitous things in this strange and fascinating world of ours. If you were to pluck two identical-appearing items off an assembly line and look them over very carefully, you would find they are not really identical after all. Even identical twins fail to live up to their label. There has probably never been a pair who were strictly identical either in physical appearance or personality. The two sides of a person's face are not even identical.

Among the more widely noted things that vary is the weather. We pay close attention to it because it affects the way we dress, what activities we can comfortably engage in, and how fast we can get from one place to another.

The rate of remuneration for a specific kind of work will vary from one geographical location to another—as will the cost of living.

Sales for any product you care to name will vary according to the time of day, the day of the week, the season of the year, geographical location, stage of the business cycle, and so forth. I could go on and on. But better yet, just look around you. I'm sure you will quickly agree that we live in a world where variation, rather than uniformity, is the natural state of things.

Of course, since this is a book about statistics, we will emphasize variation as it occurs among measured values. To illustrate: The set of numbers 3, 3, 3, 3, 3 has no variation. On the other hand, the set 1, 3, 5, 7, 9 does have variation but less than the set 1, 5, 10, 15, 300.

This book is most emphatically not a training manual for future statisticians. Still, I believe that you, a future statistical critic, maybe even a future Master Statistical Critic, would be shortchanged if it didn't encourage you to think quite hard about the question "How does one go about measuring variation?" This is not a frivolous question. Nor is it a question whose answer is self-evident.

Although many people have a pretty good feel for *averages* long before they study the subject in statistics classes, relatively few have the same visceral feel for variation. Indeed, the idea of measuring by how much several numbers differ from one another sounds to them like gobbledygook.

It is a challenge. But it isn't gobbledygook.

ROMANCING THE VARIATION

Pretend, if you will, that you live in an earlier time in a world much like our own, full of variation, but that no measure of variation has yet been "invented." Imagine that you have been asked by the mayor of your city to carry out a fact-gathering mission in nearby Tiny Town, a candidate for annexation.

The kinds of information the mayor asks you to gather are the following: (1) the number of households, (2) the average number of people per household, and (3) the amount by which the number of people per household differs from one household to another.

As you ponder your assignment, you come to realize that two parts don't seem so bad. That is, (1) requires only a count of the households and (2) calls for the calculation of an average, a skill we will assume you already possess. However, (3) could be more challenging because it requires you to measure variation. As stated above, no one, yourself included, yet knows how to measure variation.

There are two advantages to approaching variation in this admittedly farfetched way: First, it will direct your attention away from the many arcane formulas and toward the simple underlying logic of the measures. Second, it will serve to emphasize the fact that, whereas variation is inherent in life, methods of measuring variation are necessarily man-made.

Anyway, you proceed to Tiny Town and quickly learn that the city has eight households. You then go from door to door collecting information. Unfortunately, during the daytime hours, you are able to question a member of only six of the households. You find that the numbers of people in the households from which you *can* obtain information are 3, 4, 4, 4, 5, and 5. You quickly determine that, for these few households, the mean number of members is 4.17 and that the median and mode are both 4.00.

But now, what about this variation business? Chances are, the first measure of variation you would come up with is the *range.* You might reason along lines like "The largest household is made up of 5 people and the smallest, 3 people. If I calculate 5 - 3, the difference of 2 would necessarily be a measure of variation."

You would be right. The range *is* a measure of variation. It just isn't a very good one, as you learn that evening when you make a follow-up call on the two households that were inaccessible earlier in the day. You find that one of these two households has only 1 member and the other has 6. Although this new information causes the arithmetic mean to drop by only a trifling .17 to 4, it causes the range to explode to 5, more than double the preliminary value!

Clearly, for most purposes, the range is simply too sensitive to the two most extreme values in the data to serve as an effective overall measure of variation. It would be better, you reason, to develop a measure of variation which takes all of the relevant values into account.

You now pause to ponder more deeply the question, "What exactly do I mean when I say I want to measure variation?" You might come to reason as follows: *since the amount of variation is small when the numbers are very close to one another and large when they differ by a great deal, a sensible way to measure variation would seemingly be to calculate the average difference from some representative value, like the mean, for example.* That is, you could determine by how much 1 (the size of the smallest household) differs from 4 (the mean of all 8 households), by how much 3 differs from 4, by how much 4 differs from 4, and so forth. Once you know all these individual

differences, or *deviations* as they are called, you can average them. If the resulting average is large, that means there is a lot of variation; if it is small, that means little variation.

That makes sense. Alas, you put the idea to work with results as shown at the bottom of Column (3), Table 5-1.

That is, you find that the sum of the differences is zero! Zilch! Zap! Maybe it's a fluke, you wonder. You try the method on several other sets of numbers and discover that the sum is *always* zero. Dang!

But wait a minute! What if you try averaging the differences, while treating them as if they were all positive? If you throw away the minus signs appearing in Column (3) and add the resulting 3,1, 0, etc., you get a total of 8. Dividing this total by the 8 households gives you 1. That would mean: The *number of people per household in Tiny Town differs from the mean of all the households by an average of 1 person.* A mouthful to be sure. Still, it *does* make some sense.

The *mean deviation,* as this measure is called, possesses much intuitive appeal and has found some useful applications in descriptive statistical analysis. For most purposes, however, use of one of the measures described next is preferred.

Because you are interested only in the *magnitudes* of the differences from the arithmetic mean and not in their *signs,* you soon find that you can accomplish much the same thing as was accomplished with absolute differences if you resort to (1) squaring the differences from the mean, and then (2) averaging those squared differences. The resulting measure, if you were to stop at that point, is the *variance.* Calculation of the variance for our household-size data is demonstrated in Table 5-2.

Notice that the variance (computed here as a purely descriptive measure*) is 16/8 = 2.00. The variance is an important measure in its own right. However, since it is

*Most inductive applications require that the denominator of the variance and the standard deviation be reduced by one.

Table 5-1 A Common-Sense Stab At Measuring Variation

	(1)	(2)	(3)
	Number In Household	Average Of All Households	Difference (1) – (2)
	1	4	–3
	3	4	– 1
	4	4	0
	4	4	0
	4	4	0
	5	4	1
	5	4	1
	6	4	2
Total	—	—	0

difficult to express what is being measured by an average of *squared* differences, often the positive square root of the variance will be used instead. That square root is the *standard deviation,* which has the virtue of being expressed in the same units of measurement as the original data, that is, in minutes, dollars, pounds, or, as in our example, number of people.

That is not the standard deviation's only virtue. Another is one which statisticians have made good use of in a variety of ways. For a great many kinds of data, about two-thirds of the observations will lie within an interval bounded by the arithmetic mean, plus and minus the standard deviation. Suppose that after you calculated the mean of 4 you asked yourself whether 4 is really a very good representative value. Let us say, for the moment, that you found the standard deviation to be 3. That would mean that approximately two-

Table 5-2. How Work Is Done To Obtain Variance Or Standard Deviation

	(1)	(2)	(3)	(4)
	Number In Household	Average	Differences (1) − (2)	Squared Differences
	1	4	–3	9
	3	4	–1	1
	4	4	0	0
	4	4	0	0
	4	4	0	0
	5	4	1	1
	5	4	1	1
	6	4	2	4
Total	–	–	–	16

Variance = 16/8 = 2.00; Standard Deviation = $\sqrt{2.00}$ = 1.41

thirds of the household-size values should fall between 1 and 7 (4 ± 3). You might have good reason to wonder whether the average of 4 is worth very much as a representative value. (Of course, with so much variation in the data, you would have similar reservations about any other number.)

On the other hand, suppose the standard deviation was found to be .1, making an interval of 3.9 to 4.1 (4 ± .1).

In such a case, 4 could be employed as a representative value with utmost peace of mind.

In our household-size example, the truth, as the saying goes, is somewhere in between. The variance was found to be 2 and its square root 1.41. Thus, we have 4.00 ± 1.41 which is equal to 2.59 to 5.41. Are two-thirds of our values really within this interval? Alas, no. The scantiness and discreteness of our data work against perfection. Nevertheless, six of the eight values *are* between 2.59 and 5.41. That's 75 percent; close enough, perhaps, to suggest that the two-thirds rule is pretty dependable.

The above remarks have shown that there is more than one way to measure variation. They have also shown that the most important and most widely used ways are related in some manner to the concept of averaging differences from the arithmetic mean. It has also attempted to give you insight into why a small amount of variation is usually preferred to a larger amount. But small or large, it is better to have some sense of the amount of variation than to be ignorant of it.

OUT OF SIGHT MAYBE, BUT NOT OUT OF MIND

As a statistical critic, you are not likely to have to remember, in any detail, the several measures of variation described in the preceding section. It is even possible that you will live out your life never having to calculate a standard deviation or even a range.

But please don't interpret these remarks as suggesting that you can ignore variation altogether. As emphasized earlier in this chapter, we live in a world replete with things that vary; ignoring that fact will neither make variation go away nor render it inconsequential. Moreover, there are bound to be times when taking variation into account will spell the difference between accepting and rejecting a false idea.

Let us contemplate a simple example: Suppose that Fred is faced with having to take a course in statistics some time soon. Assume that he comes to you for advice about which of two professors—let's call them Professor Frank and Professor Stein—to choose for the class. (He will be thrilled to get out of the course with a C.)

A little research on Fred's behalf reveals that the average grade given by each of these professors over the years is C. Would it then be a matter of indifference which professor he studied under? Not necessarily.

Suppose that Professor Frank has seldom given a grade other than C (in other words, there is little variation around his long-run average of C) and that Professor Stein has seldom given a grade other than A and F, and these she has given in about equal numbers (bimodal data revealing that her long-run average of C could be misleading).

Professor Frank, who apparently judges the great majority of students to be just so-so, would be the natural choice. Granted, getting a grade better than C would be difficult, but so would getting a grade below C.

Professor Stein, who seemingly views students as either brilliant or stupid, should be avoided, even though an A grade would be a possibility. To repeat the principal point of this section: Sometimes an average by itself just doesn't provide enough information to help with rational decision making. Often we need to know whether the observations tend to be close to the mean or far away from it.

Examples of faulty conclusions resulting from ignoring variation are fairly plentiful. For example, did you ever hear the one about:

- the man who drowned while wading through a river whose depth was only two feet, on average?
- the squadron of soldiers who showed symptoms of malnutrition even though all the soldiers on this same base were very well fed, on average?
- the disappointed gardener who failed to make his small vegetable garden flourish even though the soil on his two-acre lot had, on average, all the right nutrients?
- the man who [groan!] sat with one leg in the oven and the other leg in the freezer and, on average, was very comfortable?

A former student of mine attended a meeting where the feasibility of establishing a new bank in a nearby suburb was discussed. The main spokesperson for those in favor of a new bank pointed out that the average annual increase in population in that suburb over the most recent ten years was about 500 people—plenty, he asserted, to justify another bank in the community.

What he failed to mention was that, ten years ago, the rate of immigration of people into this suburb was about 3500 a year and that the rate of annual immigration had declined to a mere 50 people, give or take a few, during the most recent four years. The average over the ten-year period might well have been about 500 people, just as this proponent claimed, but a single year was contributing disproportionately to that average—and that year was the one furthest removed from the time of the discussion.

Sometimes a single observation is compared with an average with rather dramatic results—dramatic, that is, as long as no mention is made of variation. Here is an example to think about, admittedly contrived and yet similar in important ways to many situations found in real life:

An acquaintance boasts to you that he can do two more push-ups than the average of all other students in his physical education class. What has he really told you? Only that he is above the class average in this respect. He certainly hasn't said that he is the best in the class, or even one of the best, but his boast could easily be interpreted in that way.

Without some knowledge of the amount of variation around the average, you can't even make a shrewd guess as to where he stands in relation to the others. For example, it is possible that your acquaintance is not even in the top half of the class!

If the modal number of push-ups were very high and the arithmetic mean were pulled down by several students who could do no push-ups at all, your acquaintance might be able to do two more push-ups than the class average and still be below the majority of class members. Granted, such a situation is unlikely. *However, a large standard deviation in the number of push-ups for the class overall isn't necessarily unlikely at all.* In such a case, two over the mean would not be particularly noteworthy.

The manufacturer of a new economy car claimed that the introduction of this car had converted many used car buyers into new car buyers because one ecology minded movie star had reported getting 40 miles to the gallon of gasoline. This mileage, the manufacturer asserted, greatly surpassed the mileage achieved by the average used car.

Clearly, the propriety of comparing the mileage achieved by *one unusually economical* car of this make with the average of all used cars is questionable. (Even comparing it with the average of all cars of this same make wouldn't be very good in the absence of information about variation, however measured.)

A cigarette company advertised that one of its brands had been found by a U.S. Government study to be 43 percent lower in tar than the average of all other menthol cigarettes. Just how reassured should a smoker of this brand be? What else does he or she need to know?

We complete our trilogy of basic descriptive measures in the next chapter, where we examine the *percent* in its many forms.

6

PUFFING UP A POINT WITH PERCENTS

> Baseball is 90 percent mental. The other half is physical.
> —Yogi Berra

Next to charts, probably no statistical tool has been used more often, more effectively, or in a wider variety of ways over the years to prop up wobbly arguments than the percent. Practically any situation can be quickly improved or worsened—on paper—if one only knows how to convert drab numbers into resplendent percents.

It is also probably safe to say that few, if any, statistical measures surpass the percent in the frequency with which honest mistakes are made. The existence of both honest misuses and wanton abuses seems all the more regrettable when we realize that a percent looks magnificently respectable. The percent sign itself (%) almost seems to be standing guard, making sure the purity of the nearby number is not compromised.

Alas, the first thing the budding statistical critic must learn is that this particular guardian can be a false friend. In this chapter we survey a considerable number of ways that percents get misused, both accidently and intentionally.

SOME MESSY DISTINCTIONS

One reason why percents get misused is that much confusion exists over the meanings of "percent," "percent change," and "percent points of change." Distinguishing among these similar sounding terms will be facilitated by an example:

Suppose that on July 1, 1991, you bought 100 shares of a common stock—let's call it South Sea Bubbles—for $20 a share. Suppose also that you bought this particular stock for its growth potential. In fact, so sure have you been of its growth possibilities that you have checked the stock's market price only once a year since the day you bought it. You have made this check, let us say, on July 1, or the nearest trading date, of each year from 1992 through 1996. Finally, suppose that the stock's market prices have been as shown in Table 6-1. Now let us see what kind of sense we can make out of the sometimes tricky terminology of percents.

PERCENT

One way of keeping track of how favorably or unfavorably the July 1 market prices compare with your 1991 purchase price would be to convert the prices in Table 6-1 into what are called *percent relatives.* That is done by expressing the $20, $30, $35, etc., as percents of the base value of $20—that is, by dividing each number in the list by $20 and multiplying the resulting quotients by 100.

The percent for 1992, for example, would be obtained by dividing the market price of $30 by the base value of $20 and multiplying the resulting 1.5 by 100, giving you 150 percent. If you were to follow this same procedure for each market price in the list, the results would be as shown in the right-

Table 6-1. Market Price Per Share Of South Sea Bubbles On July 1 Of Each Year

YEAR	PRICE
1991	$20
1992	30
1993	35
1994	40
1995	30
1996	20

hand column of Table 6-2. Figures in that column tell you at a glance that the market price in mid-1992 was 150 percent of the market price in mid-1991, the market price in mid-1993 was 175 percent of that of mid-1991, and so forth.

PERCENT CHANGE

The *percent change*, like the percent relative just discussed, is a ratio with a base-period denominator. Unlike the percent relative, however, the numerator is the *change* between two points in time. Hence, the percent change in price over a year's time (mid-1991 to mid-1992) was $[(30 - 20)/20] \bullet 100 = 50$ %, the

Table 6-2. Market Price And Percent Relative Of A Share Of South Sea Bubbles On July 1 Of Each Year (Base Year: 1991)

YEAR	PRICE	PERCENT RELATIVE
1991	$20	100
1992	30	150
1993	35	175
1994	40	200
1995	30	150
1996	20	100

percent change from mid-1991 to mid-1993 was [(35 - 20)/20]•100 = 75 %, and so on.

We are now in a position to note two peculiarities of the percent change. First, percent changes are not reversible. Notice, for example, that between 1991 and 1994 the market price of the stock doubled, climbing from $20 to $40. Then between 1994 and 1996, it dropped right back to the $20 you paid for it (an event casting serious doubt on its merits as a growth stock). The increase amounted to 100 percent: [(40 - 20)/20]•100 = 100 %. However, the decline back to $20 was only 50 percent: [(20 - 40)/40]•100 = -50%.

Suppose that the appraised value of your house went from $200,000 to $210,000 during a particular year.

That increase would be five percent. However, if the next year was one of extremely adverse market conditions and the appraised value dropped back to $200,000, you could, with effort, wrest a tiny amount of comfort out of the fact that the drop was only 4.8 percent.

A second peculiarity of the percent change is that decreases greater than 100 percent are not possible in the very common situations where the original values can only be positive. However, slips happen all the time. For example, a magazine once stated that Mao-Tse Tung had slashed salaries of some Chinese government officials by 300 percent! The magazine later confessed that the figure should have been 66.7 percent. Another magazine stated that early investors in a certain technology company had lost several hundred percent of their potential purchasing power over 23 years. After some gentle teasing by a rival magazine, the figure was corrected to a still dismal 95 percent.

PERCENT POINTS OF CHANGE

Returning to our common stock prices, we note that the percent relative for 1992 was 150 and for 1993 it was 175. The *percent points of change*, therefore, was 25, a figure obtained simply by subtracting the smaller percent relative from the larger. It would be wrong to say the percent increase was 25. The percent increase was actually $[(35 - 30)/30] \cdot 100 = 16.7\%$. As you might suppose, a number representing percent points of change is often interpreted, incorrectly, as if it were a percent change. This is the error committed in Statement 11 of Chapter 1, where a change was claimed to be a modest 3 percent, when in truth it was a considerably higher 12 percent.

Here is another example—one close to home: A colleague of mine once testified before a public utilities commission whose task it was to evaluate a request for a rate increase. A spokesperson for the company presented the figure 4.8 percent, saying that it represented the most recent monthly reading on the rate of inflation for the entire country.

This number was obtained, the spokesperson explained, by taking the change in the Consumer Price Index since the preceding month and multiplying that change by 12 to put it on an annual-rate basis. Since the change in the Consumer Price Index had been .4, twelve times that number came to 4.8. The 4.8 was referred to as an annualized percent change.

Multiplication by 12 was justifiable enough, as long as other participants in the discussion were aware of the reason for it, but use of .4 was not. Because the Consumer Price Index had been substantially above 100, a change of .4 percent points was not the same as a percent change of .4, as my colleague explained to the commission. The true percent change was closer to .3 which, annualized, came to a lower 3.6 percent.

CARELESSNESS IN COMPUTING PERCENTS

One doesn't have to be a mathematical genius to spot errors in some of the percents people cite. For example, consider the quote from baseball great Yogi Berra which introduced this chapter. Clearly, baseball can't very well be 90 percent mental and at the same time be 50 percent physical—a total of 140 percent in a situation where anything above 100 percent would be impossible.

Just as bizarre is this remark made by the concerned mother of a high school student: "I knew my son wasn't in the top 50 percent of his class, but I had no idea he was in the bottom 50 percent!." What statements like these lack in correctness, they more than make up for in entertainment value. But below are some more subtle examples of carelessness.

COMMON BASIC ERRORS

In my stock-price example, I treated the various percent measures as if they were always calculated over time. Such is not the case. Any two numbers can be compared with respect to relative size through use of a percent of some kind. For example, suppose you wish to compare the official weights of two heavyweight boxers, A and B. A weighs in at 210 pounds and B at 225 pounds. A's weight as a percent of B's would be computed as follows: (A's weight/B's weight)•100 = (210/225)•100 = 93.3%.

If you want to know by what percent A's weight is less than B's, you can simply subtract 100 from the 93.3, giving -6.7 percent. An alternative way of getting this same result would be to proceed in a manner analogous to the calculation of a percent change. Begin by determining the difference between the two weights and use the difference as the numerator. The denominator is B's weight. Therefore, [(210 - 225)/225]•100 = -6.7 %.

As mathematical things go, such percent calculations are rather simple. Nevertheless, careless errors are made all the time. For example, an advertisement for a famous brand of paper towel stated that this brand absorbs 50 percent more than competitive brands because it is two layers thick rather than just one. The truth of the claim aside, the correct percent is 100, not 50: [Number of layers for the specific brand less Number of layers for other brands/Number of

layers for other brands]•100 = [(2 - 1)/1]•100 = 100%. I once heard someone say that women give birth to 100 percent more babies than men. Women, of course, do give birth to 100 percent of the babies born, but their lead over men is vastly greater than a mere 100 percent.

ADDING AND SUBTRACTING PERCENTS

Carelessness in computing percents often occurs when one has two or more percents and wishes to make collective sense out of them. For example, sometimes either adding or subtracting the percents in hand sounds like a good idea. Often it isn't.

Suppose that Peter, Paul, and Mary are the president, vice-president, and secretary-treasurer, respectively, of a company whose board of directors has just voted the top officers a salary increase. Peter, let us say, has been voted a five-percent increase, Paul a ten-percent increase, and Mary a 15-percent increase. Is the total salary increase received by these top officers 30 percent, the total of the three percent increases? Emphatically not. Nor is it 10 percent, the average of the three. To determine the true combined percent increase, we must know the base values. Let us assume that before the pay raises, the salaries were:

Peter	$ 700,000
Paul	200,000
Mary	50,000
Total	**$950,000**

This tells us that after the raises go into effect the salaries will be:

Peter	735,000
Paul	220,000
Mary	57,500
Total	**$1,012,500**

The combined percent increase, therefore, is 6.6 percent: [($1,012,500 - $950,000)/$950,000]•100 = 6.6 %.

Picture yourself as the president of a manufacturing company. You have received a report from the accounting department showing that production costs have increased 12 percent during the past two years. You know that the selling prices of your company's products have increased by only 8 percent during the same period. Should you conclude that profits have decreased by 4 percent?

You could be quite wrong if you do. Assume, for example, that the cost of production during the base period had been $1.00 a unit and that the selling price had been $2.00 a unit. The 12 percent increase on the $1.00 base brings the cost of production to $1.12, while the 8 percent increase on the $2.00 base brings the selling price to $2.16. Thus, profit increases from $1.00 to $1.04 per unit, a 4 percent increase rather than the presumed 4 percent decrease.

MISCHIEVOUS BASE VALUES

Some of the more colorful examples of misuses of percents arise because the choice of a base value is, in some way, improper or, at least, questionable. Perhaps the base value is extremely small or perhaps it has been handpicked to encourage an incorrect interpretation.

PERCENTS MISLEADING BECAUSE OF SMALL BASE VALUES

The classic example of a misleading percent resulting from too small a base value is the one about the marrying habits of women students at Johns Hopkins University. Some decades ago, Johns Hopkins abandoned a long-standing rule and began admitting female students. Shortly thereafter it was reported that 33-1/3 percent of the women students who enrolled had married members of the faculty. The story lost much of its punch when someone disclosed that only three women had enrolled in that first year and one had married a faculty member.

As I write this, I am anticipating a several-hundred percent increase in the income of my teenage son. Last year and in previous years, he earned virtually nothing. Recently, he took on a part-time job at McDonald's. He still won't be earning much, but the percent increase should be stupendous!

One of my former students related an interesting story about the way two local newspapers reported crime statistics for a small city near his home. According to the morning newspaper, homicides in that city had increased by 60 percent over the previous year! Wow! It sounds as if the residents were being slaughtered, en masse. The student telling the story said he had visions of the place being a ghost town before the next batch of crime statistics was released.

Fortunately, the afternoon newspaper set matters straight by reporting the actual number of homicides. It stated that homicides had increased from five to eight in a year's time. Eight, of course, does represent a 60 percent increase over five; therefore, the morning paper wasn't actually lying. Still, the 60 percent number was misleading, appearing as it did unaccompanied by the actual, rather small, number of homicides contributing to the large percent change.

When the base value is small, as it was in these examples, a small numerical increase can be expressed as a very large, and probably most impressive, percent change. Consider, for example, a numerical change of 3 between two points in time. Is 3 a large or a small change? The answer depends entirely on the size of the base value. If the base value were, say, 1000, the percent change would be a mere .3 percent. If the base value were 1, the percent change would be a startling 300 percent! Everything hinges on the size of the base value, and as we have seen, if it is extremely small, grotesque results can occur.

Here is a related warning: Beware of percent changes, or any percents for that matter, when the base values are not given. If the supplier of the information has nothing to hide, he or she will gladly reveal the actual values on which the percent calculations are based.

CHOOSING AN UNREPRESENTATIVE BASE PERIOD

Unfortunately, it is easier to say what isn't a representative base period than to say what is. However, if you will take another look at the numbers in the right-hand column of Table 6-2 and then compare them with those in the right-hand column of Table 6-3 below, you will begin to get the point of this subsection.

In Table 6-2, 1991 was used as the base year whereas, in Table 6-3, 1994 was used. Although the pattern of the movements over time is essentially the same in the two tables, there is all the difference in the world in the numbers themselves. Yes, we might argue about which is the better base value in this case, but the point I trust is clear enough: sometimes the very act of picking a base period has a profound effect on the impression the viewer comes to embrace. Let's look at a real-world example:

Earlier in this chapter, I described an experience of one of my colleagues, who testified before a public utilities commission with the responsibility of evaluating a request for a rate increase. In that example, I told how he had caught the company spokesperson in the mistake of referring to a percent point increase as if it were a percent increase. Clearly, that error could have been an honest mistake. Honest or dishonest, it mattered relatively little to the rate-increase numbers that were negotiated.

More dramatic, and more likely to have been an intentional abuse of percents, was an argument utilizing average annual percent changes in the company's earnings. The spokesperson produced figures showing that the company's average annual percent increase in earnings since a specified past year was significantly below similar earnings growth rates for a broad cross section of other companies. If my colleague had conceded that the base year chosen was an appropriate one, this argument might have proved pivotal in winning the rate increase. However, this base year was, from all appearances, one that had been carefully handpicked to make the company's earnings growth seem particularly sluggish relative to those of the other companies in the comparison.

During the year preceding the one picked as the base year, the public utility had been granted a rate increase which showed up in the form of a substantial jump in earnings the following year. When the average annual percent rate of growth was computed, using this inflated earnings figure as a base, the comparison used by the proponents did indeed show their company's earnings growth lagging behind that of most of the other companies.

My colleague, however, did some figuring of his own. He computed average annual percent rates of growth for the same collection of companies that the proponents had used, but he selected a variety of base years, spanning, but excluding the one selected by the proponents. He found that by using any other recent year for a base, the average annual percent increase in this public utility's earnings surpassed the rates of practically all of the other companies in the comparison. The public utility's spokesperson had selected the one and only year out of several recent years

Table 6-3. Market Price And Percent Relative Of A Share Of South Sea Bubbles On July 1 Of Each Year (Base Year: 1994)

Year	Price	Percent Relative
1991	$20	50.0
1992	30	75.0
1993	35	87.5
1994	40	100.0
1995	30	75.0
1996	20	50.0

that could have been used to tell the story the way the company wanted the commission to hear it.

When evaluating percent changes, it is always wise to ask yourself whether the source has an axe to grind and, if so, whether the base period is unusual in some way that will bolster the source's biased position. Make the source work very hard to convince you he/she is playing fair.

THE OPPORTUNISTIC CONSTRUCTION OF A PERCENT

We now turn to a subject whose importance to the budding statistical critic—and to any other socially conscious citizen — cannot be emphasized too strongly. I am speaking of the *opportunistic construction of a percent* (OCP).

In a nutshell, if the statistical deceiver wants to make a figure seem small, he or she can express it as a percent of something large. Alternatively, if the statistical deceiver wants to make a figure seem large, he or she can express it as a percent of something small. As mentioned in Chapter 1, this technique is widely used—usually in connection with socially important and highly emotion-charged subjects. In this subsection we will examine several examples of this sometimes far-reaching form of statistical mischief.

You might recall that in Chapter 1, I asked you to respond to the following statement, one much like those frequently heard when illegal drugs are discussed: "If I were to learn that ninety percent of heroin users had, before developing the heroin habit, used marijuana regularly, I could properly conclude that marijuana use will likely lead to heroin use." At that time, I asked you to imagine three groups of people A, B, and C. Group A is very large and is made up of those people who use marijuana regularly.

Group B is much smaller and is made up of those people who use heroin regularly. Finally, Group C is smaller yet and is made up of those people who previously used marijuana regularly and then went on to use heroin regularly. We will reflect upon this statement a second time here, but this time in a way less taxing to the imagination.

In Figures 6-1 and 6-2, Group A is represented by a large square—bold, if it is the square we are focusing on at the time, and pale, if our attention is being directed elsewhere. Group B is represented by a smaller square and it is also either bold or pale, depending on whether it is a square we are concentrating on at the time. Group C is represented by the smallest square and is always clear. The spotlight near the director's right shoulder will help you keep your attention focused on the right places.

The statement presented in Chapter 1 and repeated above is illustrated in Figure 6-1. Here, remember, we really care about Groups B and C, but must not forget about Group A altogether. Group C is a large percent of the relatively small Group B, which is to say, "A large percent of regular heroin users were previously regular marijuana users."

But that doesn't bother some people who prefer to point out that, overwhelmingly, most regular marijuana users never become heroin users. Their position is illustrated in Figure 6-2, where Group A is presented in a manner saying that it is really important and Group B is played down. When we express Group C as a percent of Group A the resulting percent is small.

So who is right, the people who say that marijuana use leads to heroin use or the people who say it doesn't? If you were to ask me to take sides, I would have to say that I couldn't do so on the basis of the two pieces of evidence we have been given. The flaw in both arguments is that simple percents are being used to establish—or argue against—assertions about a cause-and-effect link between the two drugs. As we will see in Chapter 11, establishing a causal connection in this realm, or just about any other, is very difficult. It isn't even easy to spell out exactly what we mean by "causal connection."

For now, suffice to say that it isn't good practice to seize upon a percent, just because it is close to 100 (or 0), and embrace it as proof of something. Whether a percent is large or small depends importantly upon its base value. There will always be people who find that a large base value helps their cause while others find a small base value more to their liking.

Let us contemplate another statement from Chapter 1: "If I were to learn that ninety percent of the murders involving guns in this country were committed by normally law-abiding citizens, either by accident or in a moment of great emotional trauma, I would tend to support stricter gun-control laws in the expectation of reducing the number of such unfortunate occurrences."

This time, let us call Group A, "people who own guns." Group B, mercifully a much smaller group, will be "those people who have committed murder using a gun." Group C is smaller yet and represents a subset of both the other groups, "people who are normally law-abiding citizens but who, nonetheless, have committed murder using a gun." Figure 6-3 emphasizes Groups B and C, and shows how one can make a case for stricter gun-control laws on the grounds that (apparently) just having guns handy has led normally law-abiding citizens to contribute in an important way to the number of murders committed with guns.

Figure 6-4, where Groups A and C are emphasized, shows the other side of the argument. It points out that the

FIGURE 6-1

FIGURE 6-2

vast majority of gun owners never commit murder; hence, stricter gun-control laws aimed at protecting the few would represent an unnecessary encumbrance for the many.

Most of us tend to take an acquaintance's suicide threat with a grain of salt. After all, a great many people talk about committing suicide while only a small percent actually do it.

FIGURE 6-3

FIGURE 6-4

(See Figure 6-5, where A represents those who threatened suicide, B those who committed suicide, and C those who committed suicide after threatening suicide). On the other hand, mental-health authorities have, in recent years, attempted to impress the public with the need to take such threats more seriously. Recent research has indicated that three-fourths of those actually committing suicide had previously threatened to do so. (See Figure 6-6.)

FIGURE 6-5 FIGURE 6-6

Viewers of television talk shows are often informed by celebrity guests that they (the celebrities) went through a period early in their careers of great deprivation, including near-starvation diets and the need to live in their cars. Perhaps it is true that a large percent of today's top celebrities did indeed endure such hard times. However, a person contemplating a career in show business would be ill-advised to embrace what might seem the logical conclusion: "If I am willing to endure a period of extreme privation, I will one day be a big star in show business." Following the examples in Figures 6-1 through 6-6, make sketches showing how, for this example, each of the relevant percents would be calculated.

TURNING THE TACTIC ON ITS HEAD

Many of the statistical fallacies discussed in this chapter thus far have had the same underhanded motive in common: If I have a number in hand and want to impress you with it or, better yet, deceive you with it, I might be more likely to succeed if I look around for something to divide into that number. With any luck, the result will be dramatic. This tactic often succeeds in turning a "sow's ear" of a number into a "silk purse" of a percent.

But let us not overlook the fact that there is more than one way to make a silk purse; sometimes one *begins* with a sow's ear of a percent. That is, sometimes a reversal of the usual procedure will yield big benefits.

For example, let us say that I am a fund raiser for research aimed at discovering the cause of the dreaded illness, Berenstein-Barr Disease. (Don't worry about getting it, I just made it up.) Berenstein-Barr is a neurological disorder characterized by extreme pain and progressive paralysis that eventually affects the entire body. The duration of the disease is highly variable, lasting anywhere from a few days to several years.

The sheer terribleness of the disease is a plus from my standpoint as a fund raiser; potential donors can readily identify with sufferers of such a merciless sickness. Unfortunately (for me), the disease is rather rare. In the United States, it strikes only one person in 100,000 during a lifetime; that is, 1/100,000 = .00001, which, multiplied by 100, is a mere .001 percent.

Now a percent like .001 isn't likely to throw prospective donors into a frenzy of generosity. As a spokesman for a good cause but, nonetheless, a salesman, I am forced to think about how to inject more drama into the situation. My solution? I multiply the .00001 by the total number of people in the United States (recently, an estimated 264.6 million), an exercise yielding over 2600 suffering people. Twenty-six hundred! Now there's a number I can put to good use.

Or, if I am afraid that 2600 won't serve my cause adequately, I can always multiply .00001 by the number of people in the entire world (recently estimated to be 5,423,000,000). Doing so, I arrive at well over 50,000 suffering victims. Fifty thousand, or even 2600, are numbers that sound much more impressive than the tiny percent used in their calculation. This statistical trick is called the *broad-base fallacy.* Its use is becoming increasingly common in courtrooms as DNA evidence finds its way into criminal cases. For example:

Prosecuting Attorney: The DNA evidence shows that there is only one chance in a million that the murder was committed by someone other than the defendant.

Defense Attorney: Yes, but metro Gotham City alone has 5 million residents. That means we could—without even going outside the city limits—find four people other than the defendant who could have committed the murder.

Although we have covered quite a lot of ground in this chapter, we are not yet finished with the subject of percents. Percents, and things resembling percents, will appear in later chapters. In the meantime, let us consider ways in which descriptive statistical facts get used—and sometimes abused—when people make comparisons.

7

CARELESS COMPARISONS

> **Question:** Does it cost society more to send a young person to the state pen or to Penn State?
>
> **Answer:** It costs seven times more to send someone to the state pen.
> —*What Are The Chances?**

To be human, it seems, is to make comparisons. Although probably only a minority of the comparisons we make—or that others make for us—involve statistics in an important way, those that do can be pretty tricky. Indeed, the things that can interfere with the propriety of a statistical comparison are, as they say, legion. This chapter surveys several of these.

First, though, permit me to illustrate the potential subtlety of improper comparisons by drawing upon one of my favorite "oldies but goodies." This was once my idea of a "pure play" comparison, one that couldn't possibly be improper because its very simplicity precluded it. The example is shown on the following page on the left side. My annotations, written in more innocent days, are on the right side.

*By Bernard Sisken and Jerome Staller with David Rorvick (New York: Crown Publishers, 1989), p. 106.

The National Baker's Service, maker of Hollywood Bread, claimed that Hollywood Bread had fewer calories per slice than other brands of standard commercial breads.

a slice is a slice
Same way of measuring?
Proper comparison

As you can see, I gave this claim better marks than the Federal Trade Commission (FTC) did. The FTC maintained that Hollywood Bread has as high a caloric content as any other bread—about 276 calories per 100 grams—and that the only reason a slice of Hollywood Bread contains fewer calories is that its slices are thinner.

As this chapter will show, there are any number of reasons why two things, ostensibly similar, can be accessories in a misleading comparison.

AN UNTIDY COMPARISON

Where statistical comparisons are concerned, it is axiomatic that the two (or more) statistical measures to be compared should be computed in the same way. Here is an example where subtle liberties were taken with this rule with mistaken and unfortunate results.

Chances are you have never heard of Leominster, Massachusetts, a smallish factory town about 45 miles West of Boston. Leominster existed in relative obscurity until the fateful night of March 13, 1992. That was the night that the NBC news show "20/20" brought the town to the nation's attention. The subject was autism.

The guest, a Mrs. Lori Altobelli, mother of three children the second of whom is autistic, had, it seems, discovered that the autism rate in her city was exceptionally high. The reason for this, she argued, was that parents' genes had been damaged by hazardous fumes emitted by the Foster Grant sunglass factory that operated there from the early 1930's to the mid-1980's.

Altobelli's suspicions were first aroused as a result of chance encounters with two Leominster-raised fathers of autistic children. That the three of them should all be parents of autistic children seemed to Altobelli like more than just a coincidence. Spurred on by this impression, she prevailed upon the Massachusetts Department of Public Health (MDPH) to investigate. She also began taking steps to bring the matter to the public's attention.

What is perhaps her most persuasive statistical evidence—until it is scrutinized closely—can be summarized as follows: Fewer than 9000 children live within a one-mile radius of the former Foster Grant location. One-hundred-and-four cases of autism have been discovered in this area. Thus, in round numbers, about one child in one hundred in this area is autistic. It is estimated that nationally only one child in 750 is afflicted with the disorder.

That "20/20" show caused quite a stir. Reportedly, NBC received some 300 phone calls from anxious parents or parents-to-be who had at some time been exposed to fumes from the Foster Grant factory. One such parent-to-be, a former resident of Leominster, even phoned Altobelli only to be told that a couple living one block away from where this caller grew up had three autistic children!

Could Altobelli be wrong? She could indeed. First off, some people have moved out of the Leominster area over the years. Consequently, substantially more than 9000 children had lived there over the relevant span of time, a fact arguing for a larger denominator when calculating the local autism rate. (Altobelli is apparently willing to count autistic children who no longer live in Leominster in her numerator; it would seem proper that non-autistic children who no longer live in Leominster be counted in her denominator. Also, we don't know how recent move-ins, autistic and non-autistic, were dealt with. Presumably, both such groups should be excluded from the calculations.)

To make matters still worse, there are other good reasons for questioning Altobelli's figures: Many children who had been counted as autistic by Altobelli have not been so designated by trained physicians. An independent panel of experts convened by the MDPH examined the records of 26 cases in the Leominster cluster and found that only 13 really had autism. Some children thought to be autistic actually had language disorders or degenerative nervous system diseases.

To be sure, if polluted air can lead to autism, then perhaps it can lead to other disorders of a somewhat similar nature. However, if such disorders are to be included in the local count, they should also be included when calculating the national base rate used for purposes of comparison.

When scrutinized more closely like this, the Altobelli case begins to look pretty shaky. We can sympathize with her wish to find an explanation for her own child's condition. Moreover, we can applaud anybody's research efforts to uncover the cause of this terrible handicap. Less worthy of our sympathy or applause, I would suggest, are the folks at the "20/20" show. It now seems apparent that someone associated with the show should have looked into the facts with greater care before the segment was aired.*

*See also Dana Wechsler Linden, "High Anxiety," *Forbes,* August 3, 1992, pp. 42-43.

Did you check her figures?
No. I thought you did.
Mark Hall

THE DANGLING COMPARISON

After that heavy introduction, let's relax and consider *dangling comparisons*. These are things that kind of look like comparisons but really aren't.

An advertisement states that a certain brand of pipe will deliver smoke ten to twenty degrees cooler. Cooler than what? A specific competitive brand of pipe? The average of all pipes? A smudge pot? It never gets around to saying. The ad goes on to say that this pipe will deliver up to 83 percent less tar and up to 71 percent less nicotine. Again, less than what?

Another advertisement states that a certain brand of tire will stop 25 percent quicker, but it never reveals what the tire will stop quicker than. Still another advertisement boasts that a particular brand of electric shaver has blades that are 78 percent sharper. Another says that a certain brand of denture paste will let you bite up to 35 percent harder without discomfort.

Obviously, the precise-sounding statistics are only come-ons; no real information is being conveyed.

APPLES AND ORANGES

Many things can reduce or destroy the validity of a comparison. Some are illustrated below. Vigilance and a willingness to ask probing questions are the statistical critic's best defenses.

DIFFERENCES IN DEFINITION

- In Chapter 2, I pointed out how difficult it can be to make meaningful comparisons of unemployment rates between the United States and many foreign countries. The official definition of "unemployment" used in the United States is more encompassing than that of most other countries, including as it does just about anyone who can conceivably be viewed as unemployed. Some foreign countries count only those people who register with employment exchanges.

- For many years, thousands of cases of malaria were reported in the southern United States. Comparisons of rates between the South and the entire country didn't seem to make sense. Finally, the U. S. Public Health Service launched a costly investigation to search for the cause of the "epidemic." What they learned was that "malaria" in the South was only a figure of speech used to denote "cold" or "chill."*

DIFFERENCES IN REPORTING METHODS

- Mental and nervous disease appears to be more common in men than women. Comparisons are tricky, however, because men, through their employers, are more likely to be detected and institutionalized, thereby getting their names on official records. Even in these times of high labor-force participation by women, more men make their livings on jobs for which this kind of illness would be incapacitating.
- Adult-onset diabetes appears to be on the rise, but at least some of this apparent increase is a result of better detection procedures and programs.

*Related in Darrell Huff, *How To Lie With Statistics* (New York: W. W. Norton & Co., Inc., 1954), pp. 82-83.

EQUIVOCATION

- Alice Acceptance says, "The Labor Bureau just announced that the unemployment rate is five percent." To which Jack Jaundice replies (using "unemployment" in a totally different way), "Yes, but the true rate is much higher than that! Because of union-encouraged featherbedding, many workers drawing big pay are unemployed most of the time."

- Alice Acceptance says, "The morning paper reported that arrests for prostitution are declining." To which Jack Jaundice replies (using "prostitution" in a totally different way), "Well, the real prostitutes don't get arrested. We're all more or less prostitutes: We sell our bodies and minds on boring, unfulfilling jobs."

OVERLOOKING AN IMPORTANT FACTOR

- Someone might tell you that bungee jumping is safer than lying in bed. Although it is undoubtedly true that deaths from bungee jumping are very rare, the victims tend to be people who, under more placid conditions, could look forward to many more years of life. On the other hand, many of the people who die in bed are old and already bedridden. This comparison in no way supports the relative safety of bungee jumping.

- Over the years, aspirin has been the popular whipping boy used to demonstrate the relative safety of new prescription drugs. "The number of aspirin poisonings," we have learned, "makes drug XYZ look as safe as mother's milk."

Such arguments, however, are based on a comparison of the new drug taken in prescribed dosages and aspirin taken in excessive dosages.

THE FALLACY OF THE SHEEP

A newspaper article stated that the chances of a married man becoming an alcoholic are double those of a bachelor since figures show that 66 percent of male "souses" are married men. Maybe. But the situation is muddied by the fact that roughly 75 percent of men over age twenty are married. So 75 percent of the men account for 66 percent of the alcoholics—not, it seems, such a bad record for marriage after all.

This example reminds me of the old story about the discovery being made that white sheep eat more than black sheep. Further investigation revealed why: There are more white sheep than black sheep.

The implications of this are more serious than you might suppose: Assume, for example, that a certain Sunday morning was an especially grisly one for a particular city because six of its citizens were killed in separate traffic accidents. The breakdown according to destination was, let us say, as follows:

Going Fishing	Going To Church
4	2

Also assume, for convenience, that fishing and church were the only two destinations for drivers that Sunday.

Accidents will happen, but doesn't it seem a little odd that twice as many would-be anglers were killed on this Sunday morning than would-be churchgoers? Could it be that God really does mete out punishment on those who break the

Sabbath? Hopefully, you would postpone any such conclusion until learning the breakdown of the total population of drivers in that particular locale on that particular Sunday morning:

Perhaps twice as many people on the road were going fishing as were going to church (a case, I suppose, of more black sheep than white sheep). In such a situation, it would not be too surprising to find that twice as many would-be anglers were killed. If, on the other hand, only half as many were going fishing as were going to church, then going fishing on a Sunday might indeed appear inordinately risky. (Of course, some of the greater risk could be explained by the fact that those going fishing probably had greater distances to travel.)

It has been pointed out that many more males than females are attacked by sharks. The author of an article on the subject speculated that sharks, with their keen sense of smell, are attracted to something in the chemistry of men and repelled by women. Maybe. But don't be too quick to accept this or any other causal explanation.

What if I were to tell you that more little girls than little boys are injured in ballet classes? Would that suggest that boys enjoy greater suppleness and stronger bones and joints? I don't think so. How many little girls do you know who take ballet lessons? How many little boys? Of course injuries will be more common among little girls. By the same token, there are only certain places one can be to run the risk of a shark attack. Chances are, some digging into the numbers would reveal that there are usually more men than women in those places.

Let us now consider an incident where appropriate account was taken of the number of " white sheep" versus the number of "black sheep." Channel KCNC here in Denver presented a

week-long report about dog attacks on children. On the final night, the news show presented a list of breeds most likely to attack children, according to historical records:

The top three were Chow Chow, Labrador Retriever, and German Shepherd. My immediate thought was, "Come on! These are popular dogs. Of course they will have larger numbers associated with them, regardless of what is being studied." I was reassured when newscaster, Bill Stuart, presented some additional information regarding the relative popularity of the various breeds: Whereas the Labrador Retriever, German Shepherd, and Chow Chow were Number 1, 2, and 25 on the Popularity List, respectively, they were Number 2, 3, and 1 on the Frequency-Of-Attack List, respectively. In other words, for whatever reason or reasons, the Chow Chow does come across as a breed calling for special vigilance by guardians of young children.

While on this subject, I feel I must mention a fallacy closely related to the fallacy of the sheep but which, in some respects, is even more subtle. For want of a better label, I call it the *risk-your-life-and-live-longer fallacy.* Recall that in Chapter 1 you were asked to respond to this facetious twist on a familiar warning: "I appreciate the National Safety Council's informing us that most automobile fatalities occur at speeds of under 40 miles an hour because now I know that, for safety's sake, I should always drive like a bat out of hell."

The real message, of course, is that seat belts should always be worn when traveling by automobile, whatever the speed. Advice doesn't come much better than that. Often overlooked, however, is the fact that most driving is done at speeds of 40 miles per hour or less (residential areas, malls, etc.). No wonder there are more accidents (albeit less serious ones) under such circumstances; there is more opportunity for accidents to occur. Ignorance of the reason why most accidents occur at under 40 M.P.H. can lead one to think that driving fast is prudent.

Granted, that would be a pretty dumb conclusion to draw, but I am frequently surprised to find such reasoning being applied to other risky situations. These arguments usually get around to suggesting "If it sounds dangerous, it is probably safe; if it sounds safe, it is probably dangerous."

For example, a government commission report stated that "sixty-six percent of all rape and murder victims nationwide are friends or former friends or relatives of their assailants." Some news articles interpreted these results in preposterous ways similar to the following: "You are safer in a public park among strangers at night than you are at home."

This conclusion is completely wrong in spite of the carefully gathered and compiled statistics. Everyone has friends or former friends or relatives and most people spend considerable time in close proximity to some of these. It is not surprising, therefore, that most victims know their assailants. On the other hand, not many people go for strolls amid total strangers in public parks during the dark hours. I suggest we keep it that way.

CREATING A COMPARISON

Sometimes it is useful to *create* a comparison. In many fields of study, formal experimentation is used to determine whether a treatment (a new drug perhaps) has a specified effect on the subjects (people or animals) receiving it. The importance of comparisons is manifested by the nearly universal use of an experimental group and a control group. A treatment applied to one randomly assigned group, the experimental group, cannot be adjudged successful unless the members of that group have improved significantly more than the members of the randomly assigned control group, the group not receiving the treatment.

For example, at this writing, the subject of smart drugs, that is, drugs that will allegedly improve intelligence and memory, is receiving quite a lot of attention. Let us say that a new drug of that genre is to be tested. An experiment would probably be performed in the following way: I.Q. tests will be given the group destined to receive the new drug for a trial period and also to a group destined to receive a clinically ineffectual pill or capsule similar in appearance to the real thing, called a placebo. No participant in the experiment will know whether he or she is taking the real drug or the look-alike, and safeguards will even be established to ensure that the researchers themselves do not know, until after the experiment is completed, who is on the drug and who is on the placebo.

At the end of the trial period, an I.Q. test will again be given all subjects involved in the experiment. If the average score of those receiving the real drug is found to have improved impressively, a conclusion is still postponed until it is learned what happened to the average score of the control group. If the control group average is only about the same as before, or up significantly less than the experimental group average, the drug is credited with having a beneficial effect on I.Q.'s.

The control group is needed to ensure that the effects of other factors working on both groups are isolated from the true effects of the drug. What are some other such factors? It is hard to say. Slight differences in exam conditions? Time of day? Interaction with other drugs? I can guess and so can you. But, however plausible our guesses, we are unlikely to list everything that might be important. That is precisely why a control group is needed.

You might be interested to know that in such drug efficacy experiments the "placebo effect" tends to be more substantial than an outsider might suppose. According to the literature on the subject, any given experiment will result in improvement among approximately one-third of the subjects receiving the placebo. Such placebo effects have been observed in experiments on drugs designed to relieve cough, mood changes, angina pectoris, headache, seasickness, anxiety, hypertension, status asthmaticus, depression, common cold, lymphosarcoma, gastric motility, dermatitis, and pain symptoms from a variety of sources.* Moreover, one reputable medical researcher states: "Herpes simplex (cold sores, fever blisters) responds 50 percent to placebo medication. ...Going to something as clear cut as whether Rogaine applied does or does not stop or reverse hair fall, the fact that it improves hair growth in 67 percent (and the FDA agrees that it does) is not so remarkable as the fact that placebo control improves it in 50 percent."**

Can there be any question about the need to control for the several unknown factors that somehow make even a placebo appear beneficial? For that matter, can there be any question about the desirability of further study into the placebo effect itself?

As you might suppose, it is with the use of placebo-treated control groups that scientific thinking clashes most dramatically with humanitarian thinking. That is, what I am calling humanitarian thinking goes something like, "If there is some reason to think that the drug will benefit people with an unfortunate medical condition, especially if the condition is likely to lead to physical or mental deterioration or even death, how can one justify withholding the drug from such people? Even worse, perhaps, how can one justify letting them

*L. White, B. Tursky, and G. Schwartz, ed., *Placebo: Theory, Research, and Mechanisms* (New York: The Guilford Press, 1985), p. 4.

** Murray C. Zimmerman, MD, "Letters to the Editor," *Skeptical Inquirer.* July/August, 1998, p. 66.

believe that they might be receiving the drug when they are really receiving the placebo?"

The other side argues that, without a control group, it could take years to determine whether the drug is really beneficial and, even then, we could never be entirely sure. With a control group, determination of a drug's efficacy, or lack of it, can be achieved much more quickly, and then the new drug can be made available to other sufferers of the illness.

As this is being written, this long-standing controversy has recently been brought to the public's attention in connection with an experimental drug thought to have beneficial effects for sufferers of amyotrophic lateral sclerosis, more familiarly known as Lou Gehrig's disease.

Some outspoken sufferers of this disease have volunteered to participate in the experiments, but don't want a placebo. Others just want access to the drug without having to wait for the experimental results. That is all perfectly understandable. If I were already incapacitated and knew that in a year or less I would lose the use of my arms—as will many sufferers of this affliction before the experiments are over—I would be making the same arguments, as loudly as I could.

Still, harsh though it may seem, at this time there is simply no reliable alternative to the use of control groups. They are a very important part of the distinction between medical science and medical quackery.

The following chapter, "Jumping To Conclusions," overlaps this one and several others but is worthy of attention in its own right.

8

JUMPING TO CONCLUSIONS

> Once we suspect that a phenomenon exists, we generally have little trouble explaining why it exists and what it means. ... To live, it seems, is to explain, to justify, and to find coherence among diverse outcomes, characteristics, and causes.
>
> —Thomas Gilovich*

Sometimes we humans make mistakes not because we are deficient at something but because we are too good at something. The "something" to which I refer is our inclination to find order in things and to explain observed phenomena. When order exists and when explanations are actually attainable, this trait can bring enormous benefits.

On the other hand, when order is absent and the concept of "explanation" meaningless, chances are we will still find order and construct plausible explanations anyway. It seems to be the human thing to do. Moreover, once order—real or imaginary—has been found, we have little trouble coming up with reasonable explanations for its existence. For example, in the following series, would you expect a plus sign or a minus sign to occur next?

— — — — + + — — — —

**How We Know What Isn't So* (New York: The Free Press, 1991), p. 21.

Some people would say that minus is next because fully 80 percent of the observed signs are minus; thus, any new observation has a .8 probability of being minus. Such people would have a plausible justification for their choice. Others would say that plus is next because the series begins with four minuses, changes to two pluses, and then returns to four minuses; thus, a pair of pluses is to be expected next. These people too would have a plausible justification for their choice. The choices in this case are exact opposites, but both rest on sensible-sounding foundations. Perhaps there are people who could find within this set of pluses and minuses still different clues about the nature of the next sign. If so, their explanations might also impress us as plausible.

In view of my introductory paragraph, you might have had the good sense to say, "I can't determine what sign is next because I suspect that this is a random series." If that is the way you called it, you are right. I flipped a coin ten times and called a head "plus" and a tail "minus." The above series reveals nothing more than the observed sequence of heads and tails. (If you suspect my coin of being biased, you might be interested to know that the next ten flips yielded + — + — + + + + — —, a result with a slight preponderance of pluses.)

Notice: Those guessing that plus would be next were correct, though for no good reason, and are probably right now congratulating themselves on their cleverness. If you hadn't been in on the joke, do you think you would have searched for some kind of ordering system in these random data? I would bet on it. At any rate, doing so seems to be a very strong human inclination.

Now this chapter isn't about analyzing random events, or anything of the kind. However, one of its purposes is to alert you to our tendency to overexplain, to come up with explanations no matter what, and the tendency for journalists, politicians, lawyers, and others to do the same. My aim is to encourage you to be a little slower—and insist that others be a little slower—to impose explanations on phenomena which might either defy explanation or be explainable, but not necessarily by the first cause-and-effect theory that pops into your mind. Unlike gunfighting in the Old West, being a little slow on the draw is not necessarily a bad thing for one facing off with statistical evidence.

SHARK ATTACKS AND BALLET LESSONS

Let us return briefly to two examples from Chapter 7. Do you recall the one about how sharks tend to attack more men than women? It is difficult to chastise too harshly the journalist who guessed that there is something about the chemistry of men that attracts sharks and something about the chemistry of women that repels them. Nevertheless, wouldn't it have made better sense to first determine whether men have greater exposure to this particular risk? And why attribute superior physiological advantages to boys because collectively they have fewer injuries than girls in ballet classes? Off-the-cuff explanations are a dime a dozen. Be extremely wary of them.

THE EQUALLY PLAUSIBLE ALTERNATIVE CONCLUSION

Before accepting any conclusions based on statistics, a useful practice is to ask yourself whether other, equally plausible, conclusions can be reached from the same evidence. Immediately below are some simple examples related to this piece of advice. First a fact is presented. Then someone's conclusion (used here synonymously with "explanation") is offered. Finally, an equally plausible alternative conclusion is given by me. Please humor me on this; I don't claim that my alternative conclusions are very good. Most don't have to be too good to match the competition.

Claim: For the past several weeks, stock prices have been rising despite increases in long-term interest rates.

Conclusion Offered: The direction of long-term interest rates does not have any effect on the direction of stock prices.

Equally Plausible Alternative Conclusion: The direction of long-term interest rates might affect the direction of stock prices but not necessarily immediately and obviously; a recent surge in corporate earnings for example, might be masking temporarily the influence of interest rates.

Claim: In the United States, only 20 percent of the population receive dental care each year.

Conclusion offered: There must be a shortage of dentists in the United States.

Equally plausible alternative conclusion: Some people don't need dental care as often as once a year; others could use it and could get it if they sought it, but, for various reasons, don't seek it.

Claim: Seventy years after the enfranchisement of American women, only one wife in 22 casts a different vote from her husband.

Conclusion Offered: Nothing was gained by giving women the right to vote. With or without the vote, most women recognize the superior wisdom of their husbands.

Equally Plausible Alternative Conclusion: Just seventy years after the enfranchisement of American women, only one man in 22 has the courage to cast a different vote from his wife.

Claim: Several individual countries have improved their infant mortality rates more than the United States during the post-World War II period.

Conclusion Offered: The state of health care in the United States is inadequate.

Equally Plausible Alternative Conclusion: The state of health care in several other countries—probably primarily less-developed countries with poorer infant mortality rates fifty years ago—have improved markedly since the war.

Some examples for you to reflect on are presented below. Remember that your conclusion doesn't have to be any better than the one offered. It only has to be equally plausible, that is, just as well (or just as poorly) supported by the fact given.

Claim: Teenage pregnancies are increasing each year in spite of the ready availability of condoms.

Conclusion offered: The reliability of condoms must be much lower than generally believed.

Claim: Many more Scandinavians than South Africans are incapacitated each year as a result of skiing injuries.

Conclusion offered: Scandinavians in general have exceptionally brittle bones resulting from severe climatic conditions and hereditary factors, South Africans, on the other hand, are, generally speaking, free of these disadvantages.

Claim: It is known that the relationship between SAT scores and grade point averages is very weak.

Conclusion Offered: Universities should stop using SAT scores to determine which applicants are to be admitted and which rejected.

Let us now turn our attention in the direction of inductive statistics and, in particular, to one of the most important foundation stones of formal statistical analysis—probability.

9

SLAPDASH PROBABILITIES

> This branch of mathematics [probability] is the only one, I believe, in which good writers frequently get results entirely erroneous.
>
> —Charles Sanders Pierce

In Chapter 6 we discussed percents in their role as descriptive measures. In this chapter we look at percents in quite a different light as we shift our attention toward probability.

The difference between percent and probability is essentially one of certainty versus uncertainty. For example, it is one thing to say that .75, or 75 percent, of the people serving on a twelve-person jury are male and quite a different thing to say: "If one person is to be chosen at random to serve as foreperson, the probability is .75 that the one so selected will be male."

When we report that 75 percent of those on the jury are male, we are merely describing the group in a particular way; there is no uncertainty associated with our conclusion. However, when we consider the selection of one person at random, we imagine an act whose outcome is uncertain. We don't know prior to selection whether the person chosen will be male or female; we can only guess. But some guesses are better than others, and in view of the preponderance of male jurors, a guess of "male" seems sensible.

It sounds easy, doesn't it?

Probability should be easier than it is. After all, It is largely common sense. Moreover, probability calculations seldom call for any math beyond simple multiplication. And yet, miraculously, as Charles Sanders Pierce laments in the introductory quotation, probability trips up the best of writers and speakers and, occasionally, even a pretty good Master Statistical Critic. It even tripped up its first sponsor, you might say.

THE MISCALCULATION THAT STARTED IT ALL

Probability as a distinct branch of mathematics got its start from a gambling miscalculation. A Seventeenth Century French nobleman and playboy, known as the Chevalier de Méré, raised some interesting questions about games of chance. Granted, his interest was prompted more by practicality than by a pure thirst for knowledge. You see, de Méré was a gambler who thought he had discovered a surefire winning strategy for betting on dice, and for a while, he was right.

De Méré had prospered for a time by betting that he could get a six in four rolls of a single die. (A "die" is one member of a pair of dice.) He figured that, since he had an even chance of rolling any one of the six numbers on the die on the first roll, the likelihood of a six is the same as for any of the other numbers, namely 1/6. For four rolls, he reasoned—and here is where the error creeps in—the chance would be four times as good. Hence, the probability of getting a six on the four rolls should be 4/6 or 2/3.

This reasoning led de Méré to conclude that he would, in the long run, win two wagers for every one he would lose. Actually, his reasoning was flawed but, since his wealth was growing nicely, his error went undetected. (The correct probability of a win is about .52.)

The wager that gave probability its start was a variation on the preceding one. For reasons unknown to us, de Méré switched to betting that, within a sequence of 24 rolls of two dice, he could get a 12. Using the same line of reasoning as before, he decided that, because the probability of getting a 12 on one roll of two dice is 1/36 (there being 36 possible numbers that could be showing, only one of which is 12), in 24 rolls there must be a probability of 24/36 = 2/3 of getting a 12. Again, then, de Méré wins two times out of three—or so he convinces himself. In truth, however, the bet loses slightly more often than it wins, as de Méré's dwindling fortunes soon revealed.

His perplexity led de Méré to write a letter to Blaise Pascal, mathematician and philosopher. In finding an answer to de Méré's puzzle, Pascal, in collaboration with Pierre de Fermat, launched an inquiry into the theory of probability.

From this unlikely beginning, there emerged an elegant and useful discipline—one on which all of the sciences depend to one degree or another. We can only guess what might have happened if de Méré's second wager had been a winner rather than a loser. For example, had he specified 25, rather than 24 rolls of the two dice, his second bet would have won slightly more often than it lost!

PROBABILITY 101

Since probability theory had its origin in gambling games, it is not surprising that the earliest method of measuring probabilities, the *classical approach,* was one especially appropriate for gambling situations.

The classical approach defines the probability of an event as follows: If there are **a** possible outcomes favorable to the occurrence of Event **A**, and **b** possible outcomes unfavorable to the occurrence of Event **A**, and all possible outcomes are equally likely and mutually exclusive, then the probability that **A** will occur in a single trial is **a**/(**a**+**b**).

For example, if you have a coin believed to give heads and tails an equal chance of coming up on a fair flip, the probability of getting heads is obtained by first counting the sides of the coin, namely 2, and designating 2 as the denominator of a developing fraction.

The numerator is determined by counting the number of ways a *successful result* could occur, that is, the number of ways that a head could be obtained. Heads would be obtained by having the head side of the coin showing after the flip. Heads would not be obtained if tails landed up. Therefore, there is one way of getting a head and our fraction is 1/(1 + 1) = 1/2. (The term "mutually exclusive" is simply a careful way of saying that if a coin lands with the head side up, it cannot, as a result of the same flip, land with the tail side up.)

Similarly, we can compute the probability of getting a six with one roll of a single balanced die by counting the number of sides with six spots on them, namely one, and expressing this as a ratio of the number of sides altogether, or six. So, the probability of getting a six in a single roll is 1/(1 + 5) = 1/6.

THE FALLACY OF THE MATURITY OF CHANCES

Many people seem to believe that there is something akin to the law of atonement in probability. If a coin thought to have no built-in bias produces nine heads in a row, they reason, one would be wise to bet on tails next time. After all, if the ratio of tails to total flips is 1/2 in the long run, (as the classical approach seemingly suggests), then tails have some catching up to do. The fallacy in this line of reasoning is found in the fact that the coin has no memory, conscience, nor horsesense. That being so, the probability of getting a tail on the next flip is still 1/2.

This fallacy is often found in venues other than gambling houses: The contest entrant laments that she has never won anything before, so it is about her turn. Salespeople are urged not to become disheartened when a series of calls on potential buyers fails to result in a sale. The more losing calls one makes, so the reasoning goes, the closer the salesperson is to making a winning call.

And then there is this improbable story: Just before being operated on, a patient asked his surgeon to tell him frankly what his chances of survival were. "Why," replied the doctor, "Your chances are perfect. Statistics show that 99 out of 100 people who undergo this operation die, but, lucky for you, my last 99 patients all died. So, you see, you haven't a thing to worry about."

A FABULOUS WINNING STRATEGY

Speaking of flipping coins, as we were just a few paragraphs back, this might be a good time to call attention to a little different kind of gambling fallacy:

Books on betting sometimes describe in all seriousness this strategy for winning on a coin toss: Let the other person make the call because, if you call it, the odds are rather heavily against you. That's because people call heads seven out of ten times, but heads will turn up only five times out of ten. Consequently, if you let your opponent do the calling, you have a much better chance of winning.

This bit of nonsense sounds convincing at first. Clearly, though, it doesn't make any difference who makes the call or what it is; the chances remain 50-50 every time.

TOGETHERNESS PROBABILITY STYLE

Many errors occur when people try to take account of joint events. In this section I will show some right ways of handling (independent) joint events and then in the next section, some common wrong ways.

Let us consider the case of the flip of two coins. When we flip the two coins together, what happens to one coin has no effect on the other coin. Hence, we say that the events or

WHO WANTS TO CALL IT?
HE DOES!
NOT ME!
NO! OUR FATHER WHICH ART IN HEAVEN....
NO! THAT'S "BAD KARMA."

outcomes are independent. When we try to compute the probability of getting a head on each coin, for instance, we deal with what are called *joint events,* because we are contemplating the combined behavior of the two coins.

How might we go about determining the probability of getting heads on both coins? Well, we know that there are three possible end results:

Head on both coins
One head and one tail
Tail on both coins

At first glance, it might seem as if these three possible outcomes are equally likely and that, as a result, the probability of both coins coming up heads is 1/3. But that is not quite the case.

To show most clearly the fallacy of such reasoning, I will give the coins individual identities by assuming that one is a dime and the other a nickel. If we further assume that the dime lands head-up, there are still two possibilities for the nickel, namely head-up or tail-up. Of course, if we assume that the dime lands head-up, the same two possibilities still hold for the nickel. Hence, we see that a complete list of possible outcomes would be as shown below.

Dime	**Nickel**
Head	Head
Head	Tail
Tail	Head
Tail	Tail

Are these four possible outcomes equally likely? The answer has to be yes. We see that only one of the four possible joint events has both coins head-up. Thus, the probability of getting two heads on the two coins is 1/4, or the ratio of the number of favorable joint events to the total number of possible joint events. The probability of getting a head on the dime and a tail on the nickel is also 1/4. Note, however, that the probability of getting one head and one tail, with no order specified, is 1/2 because there are two ways of being successful out of the four possible outcomes.

What is the probability of having *at least* one head on the two coins? The denominator is still four. But the numerator would consist of the count of ways in which either one or two heads could be on top, namely three. So the probability is 3/4.

These probabilities could be figured more readily without enumerating all possible outcomes by employing the *multiplication rule of probability* which, for independent events, states: If two events, **A** and **B**, are independent, the probability of getting **A** and **B** together is equal to the product of their separate probabilities. Using this rule, we can determine, for example, the probability of getting head on both the dime and the nickel simply by multiplying 1/2 times 1/2, which, of course, equals 1/4.

WHEN KNOWING SOMETHING DOESN'T HELP

The story is told about a man who thought he could protect himself on plane trips by taking a harmless bomb along in his luggage. He reasoned that, whereas the odds against one person's taking a bomb aboard were high, the odds against two people doing so were surely astronomical.

A similar story concerns a sailor who put his head through a hole made in the side of a ship by an enemy ball with the intention of keeping it there for the duration of the battle. He explained that it was extremely unlikely that another ball would enter that exact same hole.

These stories, of course, are probably apocryphal and pretty farfetched at that. Still, we hear the same kind of reasoning—"slapdash-probability" reasoning, I call it—quite often. For example, a magazine article reported on a well-known golfer's spectacular achievement of sinking two holes-in-one back-to-back. The author stated that, although the odds against making the first hole-in-one were high, the odds against making the second hole-in-one were certainly astronomical.

And if you think the story of the cautious sailor is hard to believe, recall from Chapter 1 the advice actually given battlefield soldiers during World War I. They were advised to get into fresh shell holes because it is highly unlikely that a second shell will land in the exact same spot.

The fallacy in all these examples can most easily be explained through use of the multiplication rule and an example from a game of chance. Think about this question: What is the probability of getting the ace of spades twice in two consecutive drawings from a well shuffled deck of playing cards, assuming that the first selection is placed back in the deck before the second drawing?

Worded this way, the problem is seen to be pretty standard. We simply multiply the probability of getting ace of spades on the first drawing, 1/52, by the probability of getting ace of spades on the second drawing, again 1/52. That is, 1/52 times 1/52 equals 1/2704.

But now, suppose we have already drawn the ace of spades, replaced it in the deck, and are now contemplating the outcome of the next drawing. The slapdash answer (that is, the hasty, unthought-out answer) is still 1/2704. Unfortunately, in probability, the slapdash answer is frequently wrong. Since we know that we got the ace of spades on the first drawing, the probability of that is now 1.0, indicating absolute certainty. The multiplication rule, applied under these circumstances, reveals the probability of getting the ace of spades on the two drawings to be 1.0 times 1/52 equals 1/52— exactly the same probability of getting the ace of spades on any single drawing, with or without a prior success.

The same reasoning applies to the above examples as well. The man who takes a bomb onto a plane knows with certainty that he has it; consequently, he has not reduced the probability of there being another, more lethal, bomb aboard. The sailor who stuck his head through the hole in the side of a ship and the soldiers who got into fresh shell holes failed to realize that, although it is improbable for a missile to hit in any specified location, once it has done so, the probability associated with the known hit is 1.0; thus there is no reduction in the probability that another shell will hit in the same spot.

Similarly, although two holes-in-one, back-to-back, would be extremely unlikely when contemplated before the fact, once a hole-in-one has been achieved, that is now certain. Assuming independence, the next hole-in-one has the same probability of happening as had the first.

Finally, I once heard a religious speaker claim, “It is no more improbable that we will live again [after our mortal deaths] than it is improbable that we are living in the first place.” Granted, the fact that we are living at all is awesome —and presumably throughout much of the world’s development, even improbable. Still, the very reality of our existence means the associated probability is 1.0. Contrary to the speaker's superficially reasonable—but slapdash—remark, the probability of our living again is certainly lower than the probability that we are living now.

COINCIDENCES: MUCH ADO ABOUT VERY LITTLE

Hank Giclas is a colleague and friend of mine. His office used to be adjacent to mine in a six-office pod. Take a moment to ponder Hank's last name: Giclas. Chances are you have never known anyone with that surname. I certainly hadn't. The uncommonness of his surname once led me to ask Hank about its origin and meaning. He told me that the name only goes back a few generations and that, to his knowledge, it doesn't mean anything in any language.

I mention this because, while writing the section on "Togetherness Probability Style," I had a strange and chilling experience related to Hank's last name. Recall that the word "nickel" appears in that section several times. At one point while typing the material into my trusty Macintosh, I apparently misspelled the word, transposing the "e" and "l."

When I had finished writing that section, I, routinely turned my spelling checker loose on it. Of course, it informed me that I needed to correct the word. As always, the spelling checker also served up some alternative similar words. However, this time it did so with a difference. At the top of the list of alternative words was "Giclas."

I immediately invited Hank into my office to witness the miracle for himself. Although Hank is an expert with computers, we both just sort of stood there, jaws at half mast, not knowing what to say or think.

Later reflection has led me to guess that at some time, since forgotten, I wrote "Giclas" on something I was working on at the computer. Apparently, I unwittingly entered "Giclas" in the dictionary of my word processor. (It would have to be unwittingly because I wouldn't know how to do it intentionally.) I am willing to give my spelling checker credit for having a sensible "reason" for making a connection between "nickel" and "Giclas," though, for the life of me I don't know what that reason could be.

I realize that this incident is not one of monumental importance. Compared with, say, the midair collision of two 747s, it is hardly worth mentioning. Indeed, you, the reader, may find this event a good deal less interesting than I do, since the impact of a coincidence tends to be greatest on its participants. Nevertheless, things like this happen to all of us at some time and leave us wondering whether there isn't some divine guiding hand, or a supernatural prankster, behind them.

Since you have been such a good audience, permit me one more personal anecdote, and then we will look at coincidences from the vantage point of probability theory. At the beginning of summer quarter, 1984, I passed out a syllabus to one of my classes. The syllabus was more detailed than usual in that it indicated what material was to be covered during each class period throughout the quarter.

When the quarter got underway, a sharp-eyed student pointed out that no mention had been made of Monday, July 23, and wished to know whether the omission indicated a planned day off. I assured the student that the omission was merely human error and that class would be held that day as usual. However, when Monday, July 23, arrived, I found myself with a severe case of laryngitis and unable to hold class. Thus, the one day that had inadvertently been omitted from the syllabus turned out to be the one and only day that quarter when class was cancelled.

In view of what has been said about joint probabilities, we might find it tempting to call this an event of sufficient rarity to be quite remarkable. Seemingly, we ought to be entitled to multiply the probability of leaving July 23 off the syllabus by the probability of having acute laryngitis on July 23 and get a probability of 1 over some very large number.

But enough of that. The thing that makes coincidences less remarkable after some cool-headed reflection is that they have to do with *things that have already happened.*

So many things can happen and some combinations of things strike us as strange. Strange or not, if they have already happened, they are like the already-selected ace of spades discussed in the preceding section—completely devoid of uncertainty. Therefore, it is improper and potentially misleading to apply probabilistic reasoning and conclude that "The probability of this happening is so low that it just has to mean something." Probability has to do with things that *might happen,* not with things that *have happened.*

One of the more treacherous things about coincidences is their potential for misleading us with respect to cause and effect. For example, you try a new, highly advertised, quite expensive, complexion cream and, over a period of time, your complexion shows some improvement. Is that the result of the cream, as it appears, or something else? If you are in your late teens, age may be doing something for you that the cream only appears to do. If you only knew for sure, you might question the wisdom of making further purchases of the cream. (We pursue this topic further in Chapter 11.)

PROBABILITIES NOT WHAT THEY SEEM

Believe it or not, the classified ad in Figure 9-1, at the top of the following page appeared in a magazine that is distributed internationally.

Clearly, this famous psychic could guess right only half the time—no better than you or I could do—and still make $5.00 per guess, on the average. If he guesses right, he pockets the $10.00. If he guesses wrong, he returns the $10.00, or so he promises.

Boy or girl? What will baby be?

Famous psychic can tell from a snapshot or photo of full front view of mother-to-be. Must be after three months of pregnancy. Money back guarantee. Send $10.00 with photo to

[Name and box number given]

FIGURE 9-1

This is an example of one way that shysters can take advantage of others through probability. Another way is through sucker bets. As was demonstrated by the experience of the Chevalier de Méré, described at the beginning of the chapter, the correct probability of an event is not always what common sense suggests.

Shysters, as shysters will, have developed a number of wagers to exploit this particular failing of common sense. These wagers have a way of sounding as if the "operator" faces losing odds when, in reality, it is the "pigeon" who must lose in the long run. That, after all, is why such wagers are called "sucker bets." Because they illustrate ways that fallacious thinking is done about probability and because I would like to see you get reimbursed for the price of this book, I will share with you two of my favorite sucker bets.

The three-card come-on. The operator shows the pigeon three cards. One is white on both sides; one is red on both sides; and one is white on one side and red on the other. The operator mixes the cards thoroughly in a hat and lets the pigeon select one at random, instructing him to place it flat on the table top without looking at, or showing the operator, the selected card's down side.

If the up side turns out to be red, for example, the operator will say something like, "It's obvious that this is not the white-white card. It must be one of the other two, so the reverse side can be either red or white. Even so, I'll give you a break. I'll bet you a dollar against your 75 cents that the other side is also red." It sounds like a fair enough bet. After all, doesn't the pigeon have a 50-50 chance of winning? Wrong.

The catch is that there are not two possible cases, as it appears, but three, and the three are equally likely. In one case, the other side is white. In both the other cases, it is red, since the side showing could be either side of the red-red card. Therefore, the odds are two-to-one that the other side is red, and the operator loses a dollar only half as often as he wins 75 cents.

Unhappy birthday. Another golden oldie is the Coinciding Birthdays Ploy. This one requires a gathering of thirty people or more, but, preferably, not too many more. The operator announces, "I'll bet there are two people here with the same birthday." The pigeon, quickly calculating that the odds are about 11 to 1 against the occurrence of such an event, loses no time in taking the challenge. The pigeon has probably reasoned that there are 365 possible birthdates and only 30 people; hence, the probability is 30 in 365, or about one in 12. In truth, the odds are better than two-to-one that at least two of the thirty people will have the same birthday.

To see where common sense runs aground, you might picture the 30 people lined up in a row. Number one states his birthday, and the remaining 29 compare theirs with his. If there are no matches, number two announces her birthday and the remaining 28 compare their birthdays with hers. And so it goes. Each of the thirty persons has 29 separate chances of matching his or her birthday with another's.

The typical pigeon thinks, "What are the odds that any one of the other 29 has the same birthday as mine?" He should be thinking, "What are the odds that any one of the 30 has the same birthday as any other one of the 30."

Statisticians are fond of probability for several reasons, perhaps the most important being that its use tends to lead to good samples and its absence to poor samples. The following chapter will provide a brief look at why sampling, and hence probability, matters so much to both the statistician and the statistical critic.

10

THE MYSTERIOUS SAMPLING FACTORY

> The Hooperatings is a so-called service that allegedly tells you approximately how many listeners the average radio show theoretically has. It's like taking a bite out of a roll and telling you how many poppy seeds there are in the entire country.
>
> —Fred Allen

On approaching this chapter about sampling, I am not sure whether I should be wearing my white hat or my black hat. My dilemma stems in part from the fact that countless unkind things have been said about sampling over the years, many of them untrue and based on ignorance.

The late comedian Fred Allen, quoted above, can perhaps be excused for his acerbic insinuation that anyone must be crazy who thinks it possible to examine a few things (the sample) and draw a valid conclusion about a great many such things (the population). After all, the Hooperatings were showing that Allen's radio show was losing audience to "Stop The Music," a gimmicky big-money quiz show. But a great many less threatened folks share Allen's skepticism.

Although the validity of (proper) sampling is now well established among those close to the practice, the public at large still wonders whether it is really possible to use "some" to represent "all." As will be shown, that is not such a bad kind of skepticism to harbor, except for the unfortunate fact that, by itself, it doesn't lead anywhere. It ignores the happy side of sampling.

Sampling done right boasts many advantages, the most obvious being that information gathering can be carried out much more quickly and less expensively than examination of the entire population would permit. Moreover, if the information gatherer plays his cards right, that is, if he uses proper sampling methods, he can get results just about as good as a population study could deliver. So naturally, I would like for you to finish this chapter feeling pro-sampling.

Now for the other horn of my dilemma: Let us not forget that the aim of this entire book is to help you sharpen your critical judgment regarding statistical evidence. Much statistical evidence comes to us through the sampling efforts of others and can be very bad when careless or inappropriate sampling methods are used. Such careless and inappropriate sampling methods, I say, are deserving of some ridicule in a book like this.

The statistical critic gets most of his sampling-related headaches from the following, deeply entrenched, condition:

> There are a great many ways to sample, ranging in respectability between excellent and terrible. Unfortunately, as statistical critics, we usually find ourselves at a far remove from the sampling process. Worse yet, the samplers seldom volunteer much useful information about their methods.

My frustrations with this state of affairs sometimes get translated into the following dream about an MSF (mysterious sampling factory):

This eerie edifice seems to be located in the most remote possible place on earth and to be guarded by thousands of Arnold Schwarzenegger clones, each armed with immense futuristic guns. One such Arnold makes a menacing gesture at me which I interpret to mean, "Stay on the far side of the minefield." I shout, "Why?" He shouts back, "Because this is the place where all the world's sampling is done and it's strictly top secret!"

During a shift change, I manage to engage one of the friendlier looking white-collar employees in conversation— on the far side of the minefield, that is. He informs me that all employees have top security clearances and all have taken blood oaths to keep certain aspects of their work secret. "Can you talk about the results of any of your sample studies?" I ask. "Oh yes," he answers, "We can talk about results all we like." "Well, then," I ask, "Can you talk about your sampling methods?" "All I can tell you is...**Bam!!!** Right then his head explodes from the device implanted there when he took the oath of secrecy.

Then, with a jerk, I awaken, realizing that the gruesome experience was only a dream and the MSF only a figment of my imagination. But it is no use trying to go back to sleep. For the rest of the night, I'll be staring at the ceiling and muttering, "If it is only a dream, why does it feel so much like the everyday reality of us statistical critics?"

READ ALL ABOUT IT!

What is the daytime, wideawake remedy for this sorry situation? I don't exactly have a remedy but I *do* know one thing that *won't* work: Relying on members of the news media to judge samples for us. The typical news man or woman

seems capable, if not predisposed, to glide right by the question of how sampling was done. After all, the eventual story is just as exciting, whether it is based on good facts or bad. Besides, the public doesn't know the difference, does it? The only real remedy for such neglect—not one that can be realized overnight—is to work toward raising the level of consciousness of those who bring us the news, and, in the meantime, to stay alert to the signs of poor sampling. When you detect such signs, make a fuss. I do.

Speaking of sampling and the press, It seems that, again and again, we are soothed by the words "based on a sample study." Ever so gently, we are seduced into believing the claims of unknown people because they are "based on a sample study." It is not uncommon to find this assurance in adjacent newspaper articles—one pertaining to a fine piece of scientific research and the other to a reporter's off-the-cuff efforts to elicit the opinions of conveniently available people. If we are not vigilant, we will make the same mistake made by many others; namely we will view the quality of the two reports as equally good. After all, wasn't each "based on a sample study?" Didn't I say a few lines back, that the public is a little suspicious of sampling? I believe it is. But it also seems to look upon sampling with some awe. Come to think of it, suspicion and awe isn't such a bad combination.

THINGS YOU ALREADY KNOW ABOUT SAMPLING

If someone were to ask you how much you already know about sampling, your answer might be a quick and definite "nothing." Actually you probably have had more experience with sampling than you realize.

We all engage in sampling several times a day from the moment we take our first cautious spoonful of hot cereal at breakfast to the time we surf the television channels at night to see whether there are any shows worth staying up for. You sample when you check the oil in your car. You sample when you judge whether the water is hot enough to shave with or bathe in. You might even have a pet that knows a thing or two about sampling. Has your dog or cat ever stalked away from an offered meal after eating only one bite? Clearly, sampling pervades life.

Clearly, sampling pervades life.

You see, sampling doesn't always have to be complicated to be useful. But, fun though the above examples are, most situations call for sampling that is more carefully planned and more painstakingly carried out.

A PREFERENCE FOR INFERENCE

Samples, whether well or poorly selected, serve as relatively fast, low-cost guides to arriving at *inductive inferences* Briefly stated, *induction* is the process by which one arrives at conclusions about an entire group of items from a study of particular cases. Induction, illustrated a few lines down, contrasts sharply with *deduction,* which involves drawing conclusions from propositions thought to be true. For example:

Major Premise: All men like sports.

Minor Premise: Joe is a man.

Conclusion: Joe likes sports.

Induction, on the other hand, runs from the specific to the general:

Observation: Joe, Bill, Harry, Ike, Mac, et al are men.
Observation: Joe, Bill, Harry, Ike, Mac, et al like sports
Conclusion: All men like sports.

The two modes of reasoning differ in a very fundamental way. In deduction the conclusions are unquestionably valid if the major and minor premises are true. Unfortunately, however, no new knowledge is acquired. The premises are already broader than the conclusion, a fact suggesting that one would have to possess a great amount of knowledge about the matter at hand just to set up true premises. (The major premise in my example, by the way, is certainly not true. Not all men like sports. Don't ask me how I know. I just know. Still, if you will accept my premises, you are honor bound to accept my conclusion. Joe must like sports and that's all there is to it.)

The conclusions in inductive arguments, on the other hand, are broader than the premises. If the conclusions are true, something new has been added to the sum of human knowledge. But a price is paid for this possible extension of knowledge. There can be no absolute certainty about the validity of the conclusion, even though one might be absolutely certain about the accuracy of the facts leading up to it. In short, invalid and downright misleading conclusions can be reached by the inductive process. That is why induction requires more meticulous, painstaking care than many seem willing to invest.

Statistical inference is a term used to denote the inductive process as it is used in connection with statistical research. When an investigator gathers sample data and then attempts to make estimates of population characteristics or to test hypotheses about such characteristics, he or she moves inductively toward the drawing of a statistical inference, or, put more picturesquely, toward making an inductive leap.

CUSHIONING THE INDUCTIVE LEAP

So far, we have considered only the general structure of an inductive process of reasoning. Whether the process of reasoning is good or bad, right or wrong, all inductive arguments contain (1) a collection of facts about specific items or instances, and (2) a conclusion asserting that what is true of the few is also true of the many. Now we come to the practical question of how one tells whether an inductively drawn conclusion has a reasonable chance of being true or whether it is quite likely untrue. Let us consider four simple examples:

- A doctor extracts a small amount of blood from you and concludes that your blood sugar is normal.

- A lady shopper buys a cup of strawberries on the basis of a close inspection of four individual berries in the cup.

- A quality control engineer inspects 100 clevis pins out of a batch of 50,000 and concludes that a sufficiently large proportion of the pins in the batch are satisfactory.

- A young man has been jilted by three girlfriends. all under five feet tall, and vows never to date short girls again.

Now pretend that you are asked to evaluate the quality of each of the above inductive conclusions. How would you proceed? You would probably begin by recognizing that there are two essential steps in evaluating each one: First, you must distinguish between the sample of items actually observed, the group generalized from, and the population, the larger group to which the inductive inference pertains. If the sample is representative of the population, then the inference stands a very good chance of being true. If it is not representative, then, of course, the inference runs a high risk of being wrong—which brings us to the second step in evaluating an inductively drawn conclusion, namely making up your mind about it. That is, you ask yourself (1) whether the sample constitutes good evidence in favor of the inference drawn or (2) whether there is reason to think the sample might be biased, or in some other way, misleading. Keep these two steps in mind as we consider the four hypothetical cases.

A doctor extracts a small amount of blood from you and concludes that your blood sugar is normal.

You begin your assessment of this conclusion by clearly distinguishing between sample and population. In this case, the task is easy. The sample is the small amount of blood the doctor extracted from you. The population is all the blood in your body. So far, so good. Now you ask yourself whether the sample might be biased or misleading. You decide that you may safely assume, as does your doctor, that your blood is uniform throughout your body. Hence, anything that can be said about the small amount of blood can be assumed to hold for the larger amount as well.

A lady shopper buys a cup of strawberries on the basis of a close inspection of four individual berries in the cup.

Distinguishing between sample and population is easy in this example too. The sample is the four strawberries selected for close scrutiny and the population is all the strawberries in that specific cup. Now, you ask yourself whether there is any reason to assume that the four strawberries constitute a biased or misleading source of evidence about all those in the cup. The answer, you quickly realize is maybe yes and maybe no, depending on how the four strawberries were selected. If the sample consisted of four berries found on the top of the cup, you would have good reason to doubt its representativeness. Storekeepers and berry packers are known to put the best berries on top. A careful shopper will pour a few berries out into her hand in order to see what the ones underneath look like.

To summarize: You can't really judge the adequacy of the inference in this case because you have no information to help you evaluate the representativeness of the four sample berries. See what a difference the method of sample selection can make?

A quality control engineer inspects 100 clevis pins out of a batch of 50,000 and concludes that a sufficiently large proportion of the pins in the batch are satisfactory.

We have in this example a situation much like the preceding one. To begin with, distinguishing between sample and population is quite easy. The sample is the 100 pins actually inspected; the population is the batch of 50,000 pins about which the engineer must render her judgment.

Do the 100 items adequately represent the 50,000? Naturally, you would like to know the method of sample selection. In this case, however, the issue of representativeness is a bit more clouded than it was in the strawberry example. It isn't self-evident that a quick-and-dirty sample of the 100 handiest pins would represent biased or misleading material. Much depends on a number of subtle and largely unknown factors.

For safety's sake, however, you would probably feel more comfortable learning that the engineer had chosen the sample in a manner capable of handling these subtle and unknown factors better than humans can. I am alluding to random, or probability, sampling. In brief, with random sampling the sample items are selected by means of a random mechanism and without "benefit" of human judgment. More on this in the next section.

A young man has been jilted by three girlfriends, all under five feet, and vows never to date short girls again.

In this case, although the sample is easy to identify, the population is more nebulous than in the previous examples. Anyway, you would have good reason to doubt whether the three girls in the sample were representative of all girls in this indefinitely large population. People, including short girls, are highly variable with respect to many important characteristics. A sample of only three short girls would certainly be much too small to reflect accurately the considerable variation in personality, character traits, and attractiveness in the population being judged.

A COOK'S TOUR AROUND SAMPLING

Three fundamentally different approaches to sampling can be identified, and each has many subcategories. They are probability (or random) sampling, judgment sampling, and convenience sampling. Generally speaking, statisticians have little to say about the last two, save for issuing warnings about their inherent limitations. Probability sampling, on the other hand, is the cornerstone of just about any kind of statistical analysis in which a sample is utilized to make precise guesses about a population.

The unique characteristic of all probability sampling procedures is that the selection of the sample is made according to known probabilities. This characteristic subsumes three other features: (1) a specific statistical design is followed, (2) as noted earlier, the selection of items from the population is carried out by means of a random mechanism, such as a table of random numbers, and *not through the exercise of someone's judgment,* and (3) the sampling error—that is, the difference between the result obtained from a sample and that which would have been obtained from a comparable study of the entire population—can be estimated from the sample itself.

What, then, are judgment and convenience samples? These must be touched on because, with depressing frequency, the media call samples "random" pretty much habitually, including many that are not really random at all. Consequently, the Fledgling Statistical Critic (aspiring, of course, to be a Master Statistical Critic) must be able to tell the difference.

In a judgment sample, personal judgment plays the key role in determining which population items will be included in the sample. The selection of "representative" sample items is a matter of personal conviction rather than the outcome of an impersonal random mechanism. For example, when your favorite financial channel presents a panel of, say, four expert economic forecasters, these four people constitute a judgment sample of all such experts.

A convenience sample is merely a part of the population conveniently at hand. For example, the most funny/discouraging experience I know about, where convenience sampling got taken seriously, occurred shortly before one of our presidential elections:

A widely read news magazine reported that, according to a poll conducted by a Wisconsin editor, Candidate A was favored over Candidate B by a ratio of 5:3. But, the report continued, if Candidate C had not withdrawn from the race, he would have been overwhelmingly favored.

The magazine described these results as "puzzling." (At the time, they *were* puzzling—for about five minutes.) How, you ask, was the poll conducted? It seems the Wisconsin editor had questioned eight men in a bar! There are really only two things puzzling about the results. First, they were considered worthy of space in a respectable national news magazine, and, second, readers accepted the report quite passively, thinking, I suppose, "After all, it was based on a sample study, wasn't it?".

Judgment sampling and convenience sampling have their places in research work. But they do lack the advantages, cited earlier, of probability sampling.

TAKE ONE STEP FORWARD

A few months ago I happened to tune in to a television interview with legendary advice columnist, Ann Landers. Among other things, Ann and the show's hostess (and I, in my mind) reminisced about an attention-grabbing opinion survey Ann had conducted some years earlier. Ann spoke as if she still believes that the revelations brought forth by the survey are valid. The hostess spoke as if she thought so, too. Neither mentioned that the methodology employed was Bad ("bad" with a capital "B").

Now, I am as fond of Ann Landers as I am of my own mother. Still, since Ann has apparently never caught onto the fact that she brought a research monstrosity into the world, I feel obliged to set the record straight. Otherwise, the dead-wrong impressions engendered by her study might never die out—as they must.

Ann, in her advice column, threw out the question "If you had it to do over again, would you have children?" The answer? Fully 70 percent of the some 10,000 parents writing in—an impressive sample size—said they *would not have* children if they had the opportunity to make the choice again.

Stated bluntly, the parents who took the time and put forth the effort required to write in were, by and large, an angry, disillusioned lot and not at all representative of the population of parents. (Something like that is usually the case with *self-selected sampling*, as this passive version of convenience sampling is called.) My evidence? Other polls since then—polls conducted with greater care but getting less attention—indicate that a strong majority (one respectable poll commissioned by *Newsday* said 91 percent) of parents would do it again. Ponder that for a moment: The Landers survey says 70 percent "against" while the *Newsday* poll says 91 percent "for." "Duh! They were both based on sample studies, weren't they?"

One dubious defense of the Landers poll focused on the seeming diversity of the respondents. It was argued that respondents included parents in essentially all adult age classifications, education levels, and geographical locations.

Nuts, I say. What matters most is that some people were sufficiently charged up over the topic to sit down and write a letter. These were predominantly negative. The great majority were not so charged up and never made their more positive opinions known.

No few popular books have been structured around the worthless results of self-selected sampling: For example, bestselling author, Shere Hite, prospered by turning egregiously biased sampling results into *Women and Love,* a revelatory yarn about the secret anger women, in general, harbor toward men, in general. But, hey! She distributed 100,000 questionnaires and only received back about 4500.* How representative would you guess the opinions of the 4500 were? My guess: Not very.

Self-selected sampling is sometimes also called *volunteer sampling,* a term with an apt military coloration. Imagine, if you will, a platoon of ostensibly similar soldiers standing at attention. Now imagine that the officer in charge asks for volunteers for hazardous duty; a few step forward. If you sincerely believe that those few volunteers aren't different from most of the soldiers in that platoon, including several who would have to be carried into battle kicking and screaming, then you just might want to buy some shares in my diamond mine in Lapland.

Undoubtedly, the greatest polling fiasco of all time was committed by the *Literary Digest* when predicting the winner of the presidential election of 1936. The *Literary Digest* mailed out a staggering 10,000,000 ballots and had 2,300,000 returned.

On the basis of this huge sample, the magazine confidently predicted that Alfred M. Landon would win by a comfortable margin. Instead, Franklin D. Roosevelt received over 60 percent of the popular vote, a percent representing one of the largest majorities in American presidential history.

*See also A. K. Dewdney, *200% of Nothing* (New York: John Wiley & Sons, Inc., 1993), pp. 37-39.

Monday morning quarterbacks (including, alas, this author in a previous book) have attributed this immense error to an inherent bias in the magazine's mailing list. It was widely believed that the mailing list used—drawn from lists of subscribers to the magazine and telephone and automobile owners—favored higher income voters. In the 1936 election, it was believed, there was a strong relationship between income and party preference. This, at any rate, has been the conventional wisdom.

However, Maurice C. Bryson* argues, correctly I believe, along a different line: The problem was the magazine's reliance on voluntary response. (Self-selection again. I hope you are getting the message that self-selection not only routinely gives wrong results but, sometimes, the results differ from the truth by 180 degrees.) Of the 10,000,000 ballots mailed, only 2,300,000 were returned. Those returning the ballot, says Bryson, cannot in any sense be regarded as a probability sample. "In the 1936 election, it seems clear that the minority of anti-Roosevelt voters felt more strongly about the election than did the pro-Roosevelt majority."

Should self-selected sampling ever be used? Sometimes it can't be helped. Some kinds of medical experiments, for example, can only be conducted on volunteers: In a previous book, I describe an experiment which, as it turns out, presaged today's most widely used treatment for narcotics addiction. In this experiment, scientists put addict volunteers on large daily doses of methadone to block their craving for narcotics. The program was considered a success.

**American Statistician,* November 1976, p. 184.

It is difficult to imagine how anything other than self-selection could have been used in an experiment like this. After all, the scientists couldn't force addicts to participate. Still, when interpreting the results, one must be sensitive to the possibility of bias. The kind of addict who volunteers and the kind who refrains from volunteering might be quite different in ways that would affect their responsiveness to the methadone program.

Although I know it is damning with faint praise, there is actually one tiny advantage to self-selected sampling. The alert statistical critic can usually recognize it for what it is. Whenever people are asked to write or phone in their opinions on a subject, especially if they are required to pay good money for the privilege, you know that the sample is self-selected and that the results aren't likely to be worth the eyestrain incurred from reading about them.

On a happier note, do look and listen for the words,

> "Results are based on a scientific sample with a four [or other] percent margin of error."

Granted, even with that reassuring addendum, more is concealed than revealed about the sampling methods employed. Nevertheless, terms like "scientific sample" and "margin of error" do strongly suggest that (1) a formal sampling design of some kind was employed, (2) the samplers used an impersonal means of determining who would be included in the sample, as distinguished from looking the other way (a la Landers and Hite) while the angriest people dealt themselves in, and (3) the sampling method had a probability basis. So you see, these fourteen little words really do constitute an enlightening and helpful peek into the workings of the mysterious sampling factory.

Now and then in previous chapters we have brushed up against the subjects of relationships and cause-and-effect linkages between events. In the next chapter we will deal with these important topics at greater length.

11

RELATIONSHIPS: CAUSAL AND CASUAL

> Whether the stone hits the pitcher or the pitcher hits the stone, it's going to be bad for the pitcher.
>
> —Sancho Panza

Without question, one of the more challenging tasks faced by the statistical critic (or any conscientious citizen for that matter) involves the evaluation of claimed relationships among variables. That is especially true when the claims smack of cause and effect. This chapter shows why.

In Chapter 1, I introduced you to W. G. Brownson's amazing Electro-Chemical ring. Recall that Brownson claimed his ring would cure diseases caused by acid in the blood, including Bright's Disease, diabetes, epilepsy, and so on. At that time, I also pointed out that, although the ring was made of nothing more than industrial grade iron, many purchasers swore it cured their ailments.

This is just one example among many that could be cited where a treatment, which logically should not cure one of anything, gets credit for doing something miraculous. Let us not be too hasty to believe such claims. If we probe a little more deeply, we can see just how easy it is for a fatally ill person, and most of the rest of us from time to time, to make faulty causal connections. People with serious physical ailments tend to give the wrong things credit for their (temporarily) improved health because they don't understand the usual zigzag path that fatal diseases take.

Figure 11-1 depicts such a zigzag path. Notice that even though the general direction of the line is downward, it is possible to find two points in time, A and B in the chart, between which the patient will think he has improved.

If the patient were to try a quack remedy such as the Brownson ring at Point A, he would likely attribute his improvement to the ring and, out of gratitude, give it an enthusiastic endorsement, and then, alas, suffer a setback shortly thereafter. What is the likelihood that some such quack remedy will be tried at Point A rather than at some other point? Quite good actually, since people tend to try new remedies at times when they are feeling especially bad.

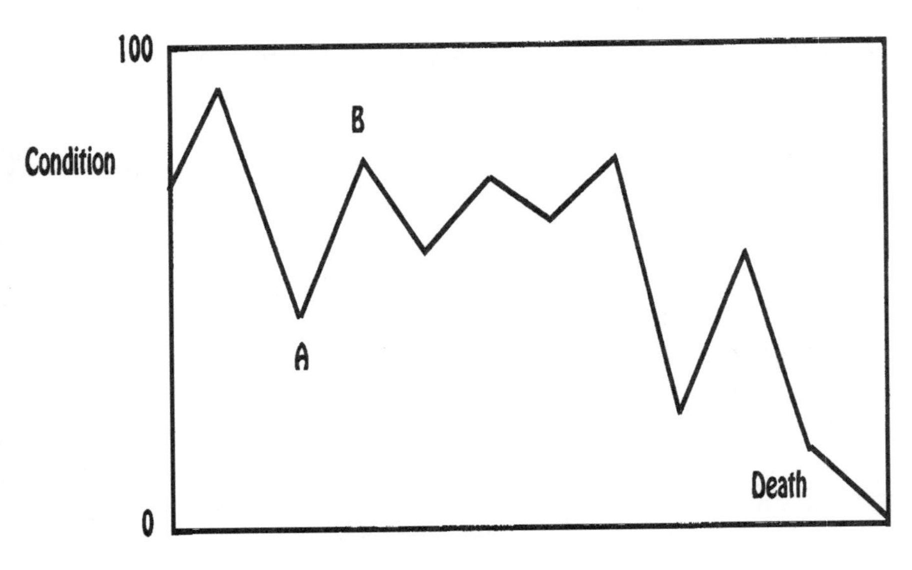

FIGURE 11-1

Keep Brownson's healing ring in mind while you ponder the following experience: Just about four years ago, I underwent surgery on my ankle to correct what my doctor called a frayed tendon. For the next two months, I was forced to hop around on a walker while the ankle mended. As often happens when one must transport his body in an unaccustomed way, I added a second injury to the first.

While learning to go up the stairs of my house, I managed to pinch a nerve in my neck! The resulting pain in my left arm was so terrible that I nearly forgot about my convalescing foot.

After some weeks of enduring this extreme discomfort, I finally saw a neurologist who checked me out and stated, in substance, "If you can continue to live with it, it will probably heal itself. There is a surgical alternative, but, in my opinion, it has more disadvantages than advantages." I respect that doctor for admitting that nature could heal me better than he could. Just as he said, the discomfort has disappeared. The whole miserable ordeal is now just a hazy memory.

But what if, in my anguish, I had selected a less honest neurologist; let's call him Dr. Fibb. Suppose that Dr. Fibb said to me "There is only one thing that can help you: The Brownson Electro-Chemical ring. Although they are in very short supply, I can get you one for only $10,000." What would have happened?

The top part of Figure 11-2 shows the likely path of the ailment with "ring therapy" absent (though this path would have been unknown to me). The bottom part of this same figure shows the path with "ring therapy" present (the only path I would have known about). Clearly, the two paths are identical and both end in recovery.

If the experience had unfolded in the manner depicted in the bottom part of Figure 11-2, would I have believed that the ring had brought about the cure? You bet. Would I have been willing to render a heartfelt testimonial on behalf of the miraculous ring? It's the least I could do. Would I have been willing to pay Dr. Fibb his $10,000 fee? I would have begged him to take it.

This error of reasoning, by the way, has an intimidating Latin name. It is called the *post hoc ergo propter hoc* fallacy,

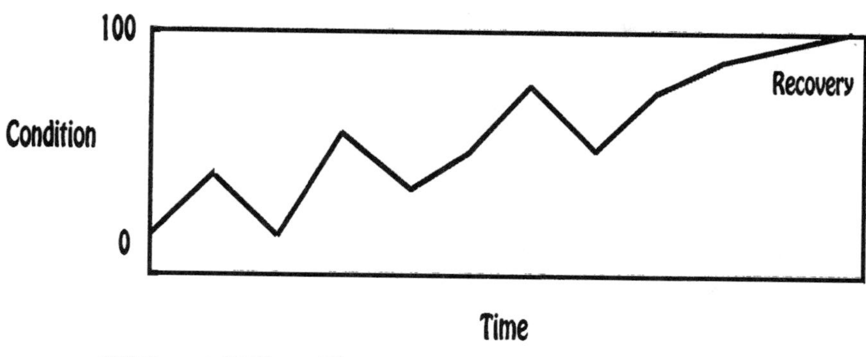

Without "Ring Therapy"

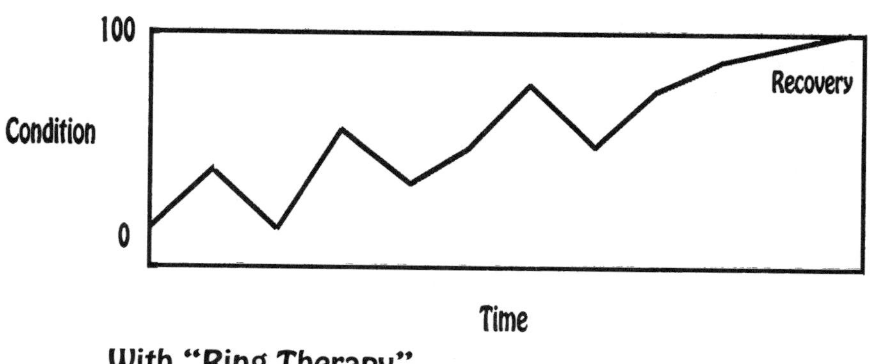

With "Ring Therapy"

FIGURE 11-2

or, more conveniently, the *post hoc fallacy*. The English translation is, essentially, "after this, therefore, because of this." Post hoc reasoning would lead one to believe that if Event **A** preceded Event **B**, then **A** must be the cause of **B**. ("I should have known it would rain; I washed my car this morning.") **A** might indeed have been the cause of **B**; but in view of the many events—some clearly relevant and some more farfetched—that also preceded **B**, one must be a little humble about declaring a "cause." I believe you can appreciate how often post hoc reasoning is grasped at by people who are sick, in pain, maybe even dying, and at the mercy of some kind of healer.

The examples about the Brownson ring are intended to illustrate how quickly we become confused when we attempt to attribute causes to events of importance to us. We all believe we know a thing or two about "cause and effect." Our vocabularies are laced with words suggesting causal linkages. Intuitively, the idea of causality seems crystal clear. Unfortunately, when we address the concept with a little more care, we often find we really don't know what we're talking about.

Permit me to confuse matters still more. Today, it is common knowledge that cigarette smoking causes lung cancer. And yet the matter has been seriously researched and hotly debated for over half a century. (Cigarette manufacturers and others with a vested interest are still resisting the conclusion.) If the concept of cause and effect is as simple as we tend to assume, why has it taken so long to establish a causal connection between smoking and this often fatal disease?

The harsh fact of the matter is that establishing cause and effect is often horrendously difficult. For example, perhaps you know someone who has been a heavy smoker for many years but who does not have lung cancer. You might even know someone who never smoked at all but who, nevertheless, does have lung cancer. Such anomalies are common and force us to abandon all hope for a simple explanation of how smoking and lung cancer are connected. Clearly, cigarette smoking must be a contributory cause. That is, it is one of perhaps several causal factors. Thus, researchers have been saddled with the task not only of establishing that smoking is indeed a causal factor but also of determining how important smoking is relative to the other contributors.

Belatedly, tobacco industry leaders came to admit that there is a statistical correlation between smoking and lung cancer but worked (and spent) like the dickens to convince the public that the relationship isn't causal. Although recent research has put the lie to the industry's main arguments and advanced our knowledge of the causal mechanism at work, there is, after fifty years, still much to learn.

Whoever said that establishing cause and effect is easy just didn't know.

CAUSE A LA CARTE

After the above discussion, it might seem futile to do further poking around with the concept of "cause," and, from some perspectives, maybe it is. From a practical standpoint, however, the idea that if **A** happens, **B** will happen, or is more likely to happen, is often a most useful one.

Maybe **B** is a crippling disease and **A** is an activity or condition somehow associated with the occurrence of that disease. If by eliminating **A**, we lessen the threat of **B**, we have achieved something worthwhile even though we have not even begun to follow the full causal web in all its far-reaching and subtle ramifications.

What follows is a menu of sorts displaying several categories of causes. Knowing them can help sharpen your thinking. That is important for the statistical critic because sloppy, off-the-cuff claims about causality are so often made in the media and elsewhere.

NECESSARY CAUSE

A *necessary cause* is a condition that must be present if an event is going to occur. An automobile must have fuel if it is to run. It won't go otherwise. In other words, if **A** is a necessary cause of **B**, then **B** will not occur without **A**.

SUFFICIENT CAUSE

A *sufficient cause* is any condition that will bring about the event alone and by itself. If **A** is a sufficient cause of **B**, then **B** will always occur when **A** occurs. If you cry every time you slice onions, then the act of slicing onions is a sufficient cause of crying. The act of slicing onions is not a necessary cause of crying because other things can also lead to crying.

NECESSARY AND SUFFICIENT CAUSE

A *necessary and sufficient cause* is any condition which will bring about the event and without which the event will not occur. That is, if **A** is a necessary and sufficient cause of **B**, then **B** will occur if and only if **A** occurs. Maybe you never cry except when you slice onions. In such a case the act of slicing onions would be a sufficient cause of crying because it will do the job by itself, but it will also be a necessary cause as well, because nothing else moves you to cry. Clearly, necessary and sufficient causes are rare.

CONTRIBUTORY CAUSE

A *contributory cause* is a factor that helps to create a total set of conditions necessary or sufficient for an effect.

Tension is thought to be a contributory cause of headaches in that it contributes to a set of conditions out of which headaches often arise. But some headaches seem to occur without tension. If **A** is a contributory cause of **B**, then **B** is more likely to occur when **A** occurs than when **A** does not occur.

Contributory causes, the kind we will emphasize throughout the remainder of this chapter, occur more often in situations concerned with the characteristics or activities of humans than do either clear-cut necessary causes or sufficient causes (and, of course, much more often than necessary and sufficient causes) and are sometimes difficult to disentangle from other contributory causes.

THE CAUSE OR JUST A CAUSE?

An error frequently made, and the source of a great many statistical fallacies, involves looking for a simple, single cause of a phenomenon which is really the result of a combination of several contributory causes. This fallacy will be illustrated with the help of an experience I had when writing a previous book. It dramatizes how easily someone can be rendered blind to all but one consideration.

It started with a report by the U.S. Bureau Of The Census stating that (in the early 1970's) the life expectancy of the U.S. born male is 67 years. This news was upsetting to many people because the figure was less than that of several other countries, especially Sweden, Japan, Czechoslovakia, and Israel. Without making any effort to do so, I accidently happened upon the remarks of four respectable writers trying to make sense out of the fact. Notice how each of these writers, let's call them Writers T,

U, N, and L, (collectively, TUNL, for tunnel vision) seizes upon a single contributory cause and treats it, with gusto, as if it were the only cause.

Writer T: Writer T wrote an article in a well-known national news magazine. He argued that the relatively short lives of U.S. males result from stress in the workplace—the pressure to perform. He essentially ignored other possible causes.

Writer U: Writer U was the author of a book critical of the quality of medicine practiced in the United States. He went to great lengths to destroy the "stress theory" before advancing the view that medicine, as practiced in this country, is nowhere near as good as it could be and should be. He essentially ignored other possible causes.

Writer N: Writer N was an authority on nutrition, so naturally, he emphasized the inadequacy of the American diet. More specifically, he presented copious facts demonstrating that Americans eat too much fat. He essentially ignored other possible causes.

Writer L: Writer L was also an authority on nutrition. However, his fixation was with the excessive carbohydrate consumption of Americans. He essentially ignored other possible causes.

I am not qualified to tell you which of these arguments is closest to the truth. However, I do feel comfortable pointing out that, if there is only one relevant cause, and if that cause is one of those noted above, three out of four of these "experts" are dead wrong! Moreover, if the one relevant cause *is not* among those noted above, then all four are dead wrong! Isn't it reasonable to think that more than one cause is at work?

A similar kind of oversimplification, but in a grosser form, is an old standard of advertising and politics. I trust you have seen before-and-after ads—the kind with a blurry picture on the left showing a girl who is clearly overweight and who also has fuzzy, unkempt hair, crooked teeth, and an unpleasant complexion. This is the "before" picture.

The "after" picture, on the other hand, is bright and clear and shows a trim girl with faultless hair, teeth, and complexion. The advertising copy calls attention to the fact that the girls are really one and the same and that the "after" version owes her new-found beauty—and resulting happiness—to such-and-such reducing plan.

Of course, the copy doesn't mention that this determined and praiseworthy girl tried a variety of reducing aids, including regular exercise, to become more slender. (The copy also fails to explain how this particular reducing plan straightened the girl's teeth and made her hair more manageable or enhanced the clarity of the picture.)

CORRELATION AND CAUSATION: YOU CAN HAVE ONE WITHOUT THE OTHER

Although the term wasn't used there, you were introduced to the idea of *correlation* in Chapter 3. Correlation analysis, along with its close cousin *regression* analysis, is the branch of statistics concerned with determining whether, and how dependably, two (or more) variables vary together.

Correlation and regression analyses are undoubtedly the most frequently used statistical techniques in social-science and business research. But whatever the application, it is necessary to draw a clear distinction between correlation and causation.

The two concepts are frequently confused, with results that can be spectacularly misleading. While it is true that correlation analysis will sometimes support a belief of the kind "X causes Y," much more often it will reveal that, "Yes, there is something going on between X and Y, but it is more complicated than you thought."

Although the following discussion will be largely free of technical tools and esoteric jargon, some use will be made of the scatter-diagram concept, a subject introduced in Chapter 3. Recall that construction of a scatter-diagram entails (1) calling one variable Y and measuring it on the vertical axis, (2) calling the other variable X and measuring it on the horizontal axis, and (3) indicating, with tiny circles, *pairs* of X and Y values within the body of the graph. A scatter-diagram based on data we will be discussing shortly is shown in Figure 11-3.

What follows are several examples involving the presence of correlation but having quite different causal infrastructures.

TWO VARIABLES, X AND Y, CORRELATE BECAUSE X IS THE CAUSE OF Y

The X variable might actually be the cause, or be intimately associated with the cause(s), of Y. For example, there is known to be a strong correlation between temperature and the frequency with which snowy tree crickets chirp in a minute. If we were to multiply the degrees Fahrenheit by

3.78 and subtract 137 from the result, we could estimate the number of chirps per minute expected from a typical cricket of this kind with considerable accuracy (as suggested by the scatter-diagram in Figure 11-3).

That the two variables are strongly correlated is attested to by the facts that (1) Figure 11-3 reveals very little variation around the mathematically fitted line and (2) the correlation coefficient (a commonly used measure of degree, or strength, of relationship, the highest possible value being ± 1.0) is an nearly perfect +.9919.

It seems infinitely more sensible to suppose that variation in the temperature somehow produces changes in the number of times our cricket chirps in a minute than to suppose the reverse. We conjure up a ludicrous picture indeed when we try to imagine a cricket driving up the temperature by going into a chirping frenzy.

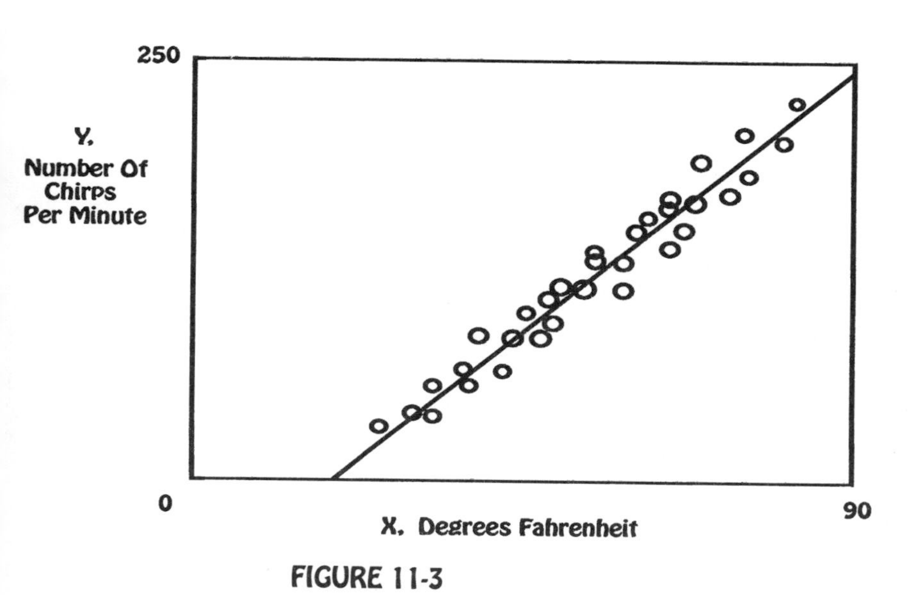

FIGURE 11-3

We conjure up a ludicrous picture indeed when we try to imagine a cricket driving up the temperature by going into a chirping frenzy.

TWO VARIABLES, X AND Y, CORRELATE BECAUSE Y IS THE CAUSE OF X

Sometimes the presumed cause is really the effect and the presumed effect really the cause. I am reminded of Chanticler, the rooster hero in the Edmond Rostand play of the same name. Early in the play, the proud bird suffers from the delusion that his crowing causes the sun to rise and the new day to begin—a clear-cut reversal of cause and effect. By the end of the play, Chanticler is more realistic. He then realizes that he has nothing to do with making the new day begin but decides to keep on crowing anyway to celebrate its arrival.

Let's say that you are a marketing researcher. A company's president asks you to find out whether the firm's sales could be increased by spending more money on advertising than it customarily has. You take the assignment and decide to determine what the experience of other firms has been. You select a random sample of businesses and run a correlation analysis of dollar sales and advertising expenditures, using the latter as the X variable and presumed cause.

Let us further suppose that you discover a pretty strong positive correlation between the two variables. What would you conclude? Should the company be spending more money on advertising? The answer depends on how confident you are that you assigned the cause-and-effect roles properly. Perhaps it is the businesses with the highest sales that are most inclined to spend heavily on advertising. In other words, there might be a causal relationship, but the chain of cause-and-effect might be running from sales to advertising rather than from advertising to sales.

TWO VARIABLES, X AND Y, CORRELATE BECAUSE THEY INTERACT

A more realistic interpretation than one dependent upon a simple one-directional cause-and-effect relationship might be that advertising and sales interact in such a way that X is sometimes cause and sometimes effect, and the same with Y. That is, maybe advertising helps bolster sales of smaller firms, which, as their sales increase, can spend more liberally on advertising. The additional advertising helps to boost sales even more, and that leads to still higher advertising expenditures. And so it goes.

TWO VARIABLES, X AND Y, CORRELATE BECAUSE OF LURKING VARIABLES

Sometimes a researcher will discover a correlation between two variables, enjoy the fact momentarily, and then become frustrated trying to figure out how to interpret it. He might begin by assuming X to be the causal variable. Alas, it doesn't make sense. He will then try Y as the cause. It still doesn't make sense. Nor does the presumption of interaction solve the mystery.

For example, a southern meteorologist reportedly found that the fall price of corn is inversely related to the severity of hay fever cases (that is, as hay fever becomes more severe, the price of corn tends to fall, and vice versa). But he couldn't convince himself that low prices aggravate hay fever or that severe cases of hay fever bring down prices.

Chances are, his problem is a common one...the nefarious *lurking variable.* A lurking variable is a variable not included in the research itself but is one which has an effect on both X and Y. The lurking variable in this case is presumably the weather. When weather conditions are conducive to a bumper crop of corn (and, hence, to lower prices), they are also conducive to a bumper crop of ragweed.

Lurking variables are annoying and confusing, but they can also be humorous. Here are some vintage examples:

- Before storks became candidates for extinction in recent years, there was reportedly a positive correlation between the number of stork nests and the number of human births in Northwestern Europe.
  This correlation, rather than adding credence to the old dodge about how babies come into the world, was simply a reflection of population growth. As population and, hence, the number of buildings increased, the number of places for storks to nest also increased.

- It is said that Hippocrates, the father of western medicine, noticed that the farther up the hillside a family resided, the better the family's health tended to be. He concluded that there is something about higher altitudes that imparts health enhancing benefits.
  Altitude might matter. However, let us not ignore the fact that the hillsides tended to be inhabited by the more affluent families, families who could afford better food and other amenities conducive to better health. In a similar vein, Dennis G. Haack* points out that endemic goiters increase in frequency with higher altitudes. But altitude does not actually cause the goiters; they result from low iodine in food, a condition more common at higher altitudes.

* *Statistical Literacy: A Guide To Interpretation* (Massachusetts: Duxbury Press, 1979), p. 212

- Shoe size and its effects on intellectual prowess seems to fascinate some statistical writers. One such author asserted that there is a strong correlation between shoe size and handwriting test scores. Another makes it shoe size and reading test scores. Yet another claims that bigger shoes are associated with higher I.Q. scores.
  Alas, wearing larger shoes won't improve your handwriting, your reading, or your I.Q. It seems that the tests were given to children ranging rather widely in age, the older ones tending to do better. Of course, the older ones tended to be larger and have commensurately larger feet.

- The prestigious financial newspaper, *Barron's,* (March 17, 1997) recently waxed playful by presenting an article based on the (actual) very strong relationship between stock prices and the consumption of Prozac and other antidepressant drugs. The central question was, "Is the sustained bull market of recent years attributable to drug-induced happy spirits of investors?" A line chart showing both variables for the years 1987 through 1996 was admittedly beguiling.
  Many things correlate across time. The proof of the pudding will come with the next bear market—if, indeed, there ever is another bear market. (Better antidepressant drugs are being developed all the time.) Consumption of antidepressant drugs should, according to the newspaper's analysis, fall as the market falls. I am not expecting that to happen.

- A dead solemn researcher some years ago reported, "Over the past several decades, there has been a strong inverse correlation between the number of mules on Missouri farms and the number of Freshmen in the nearby universities."

No, the mules were not leaving the farms and entering the universities. The phenomenon was a reflection of the increasing urbanization of the state of Missouri. With less land being farmed, there was a gradual reduction in the number of mules born. University enrollments, on the other hand, were mounting because of the demands of the changing job market.

Determining cause and effect is usually a pretty difficult task. Fortunately, the purpose of this chapter has not been to turn you into an expert on the subject. Rather, it has been to let you in on the well-kept secret that you know as much about cause and effect as almost anybody; some people just act as if they know more.

As I see it, there are two things a statistical critic should make use of in connection with cause-and-effect arguments. The first is vigilance. If you will develop the habit of paying close attention to what you are reading or watching, you will catch many questionable causal statements quite effortlessly.

The second is a "filter" based on the topics treated in this chapter: Ask yourself: (1) whether a contributory cause is being viewed as the sole cause, (2) whether what is said to be the effect is really the cause and what is said to be the cause is really the effect, (3) whether it is reasonable to think that the two variables might "take turns" being the cause, and (4) whether a lurking variable might be getting in the way of a proper interpretation of the relationship altogether. If you will develop the habit of doing these simple things, you will be a Master Statistical Critic in no time.

Contrary to its advance billing (Chapter 2, page 39), this chapter, thus far, has been strictly descriptive. That is, we have assumed that either sample or population data were available and have dwelt on the subject of causality rather than statistical inference. Such an emphasis makes sense when the reader is a future statistical critic rather than a future statistician.

Still, even a future statistical critic must be aware that, in practice, the formal analysis of relationships is often conducted on sample data with the intention of drawing correct inferences about the corresponding population. Thus, some additional cautionary points are in order. When sampling is used, such considerations as sample size and sample-selection strategy do bear importantly on the validity of the results. Unfortunately, the needed facts about such matters will seldom be given in the related newspaper article, the political speech, or the TV magazine show. (Remember the Mysterious Sampling Factory?)

Sometimes, however, such sources will give you a sense of whether the sample used was large or small. This distinction is important because, where statistical relationships are concerned, small samples can sometimes lead to bizarre results. (So can large samples—but less often.)

For example, imagine a population like the one depicted in the scatter-diagram in Figure 11-4 below. This population is totally devoid of correlation. (Though, in reality, we wouldn't know that unfortunate fact.) In such a dismal situation, it would be reasonable to expect the scatter-diagram for the corresponding sample data to look more or less just as

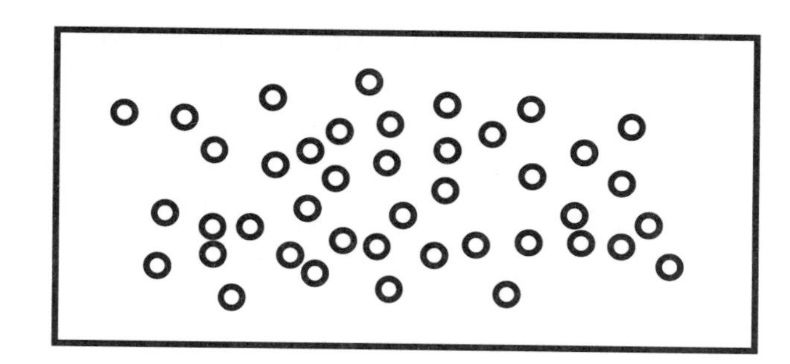

FIGURE 11-4

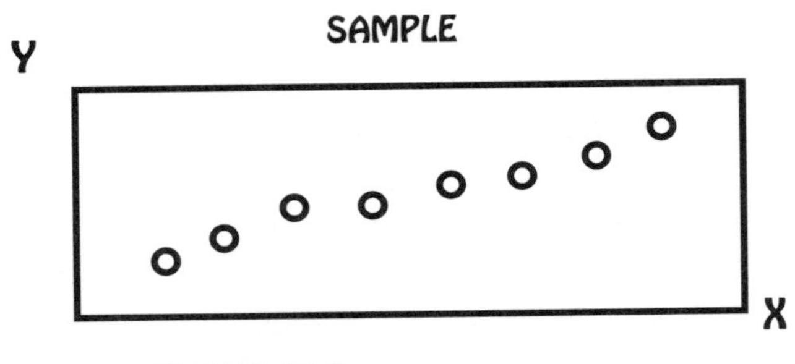

FIGURE 11-5

uninspiring. But check out Figure 11-5 above. It's loaded with correlation! Is such a thing really possible? Indeed it is; with small sample sizes, it is even fairly common. Why? No good reason. It is sometimes just the "luck of the draw," an occasional result of random sampling.

Notice that, in such a case, the sample message would be the exact opposite of that conveyed by the overall population if we had access to it. Not only that, but the sample result would be more exciting than the true population condition. People might actually talk about the sample "information," whereas the true population condition, if known, would provide little grist for polite party conversation or a newspaper headline.

Clearly, samples of the same (small) size, drawn from the exact same population, can provide very different kinds of evidence. It isn't unheard of for a correlation to "vanish" according to a second sample after the exciting results of a first sample have already made the ten o'clock news (so to speak). Be sensitive to sample size, and when you know the sample size to be very small, be skeptical of any conclusions drawn from it.

The next chapter continues to address causation, but from a rather different vantage point.

12

BEYOND PLAIN AS DAY

> Again and again a causal sequence is asserted because certain events are found [by psychoanalysts] to occur frequently in the early years of the lives of neurotics; hardly ever is there any attempt to show that these events occur less frequently or not at all in the lives of nonneurotics
>
> —H. A. Eysenck*

"It's as plain as day." What a reassuring claim! But be careful. An important aim of this chapter is to show you that "as plain as day" and "true" are not necessarily the same thing. Another important aim of this chapter is to acquaint you with a most useful perspective-stretching tool, the *contingency table.*

The basic makeup of a contingency table is shown in Table 12-1 on the following page. Notice that, not counting the locations where totals are shown, this contingency table has two rows (left-hand side) and two columns (across the top). Two rows and two columns make it a *2 x 2 contingency table*, the only size we will be discussing in this brief introduction to the subject.

*Uses and Abuses of Psychology (England: Penguin Books Ltd., 1953), p. 238.

Table 12-1. The Structure of a 2 x 2 Contingency Table

	Condition I Present	Condition I Absent	Row Total
Condition II Present	a	b	Sum a,b
Condition II Absent	c	d	Sum c,d
Column Total	Sum a,c	Sum b,d	Grand Total

Notice also that the two columns are labeled Condition I Present and Condition I Absent. If, for example, we are focusing on neurotics, as we will be in Table 12-2, then those designated "Neurotic" will constitute our Condition Present column; those designated "Not Neurotic" will inhabit our Condition Absent column. *For our purposes, the Condition Present column will always be the left one.*

Similarly, the rows are labeled Condition II Present ("From Broken Home" in Table 12-2) and Condition II Absent ("Not From Broken Home"). *For our purposes, the Condition Present row will always be the top one.*

Finally, there are four cells in the table, labeled **a**, **b**, **c**, and **d**. In future examples, numbers, rather than letters, will often occupy these cells. The numbers will represent *observed frequencies,* which are nothing more than counts of actual things, e.g., 12 neurotics from broken homes, 3 accurate guesses by a police psychic, and so forth. It is as simple as that. However, please note:

In this chapter, Cell **a** will always be situated in the table so that it represents the intersection of the Condition Present row and the Condition Present column.

AS PLAIN AS NIGHT

In the quotation introducing this chapter, Eysenck, a psychologist, shakes an angry finger at the psychoanalytic wing of the mental health establishment. We can benefit from looking closely at the substance of the scolding because, at one time or another, we all make the same mistake Eysenck accuses the psychoanalysts of making.

Eysenck discusses the relationship (if any) between broken homes and neurosis. He asserts that a psychoanalyst, upon learning of the broken home, would conclude that the condition had contributed to the patient's neurotic behavior. [It's as plain as day.] But Eysenck asks, "Aren't they the least bit curious about nonneurotics who came from broken homes? Where is their control group?"

A useful way of ensuring that we don't overlook any aspect of a question like this is to realize that the relevant people are not just those in Group **a** (Neurotic and From Broken Home; see Table 12-2). They are relevant all right, but so are the members of Group **b**, those not neurotic

Table 12-2. Basic Contingency Table For Neurosis Example

	Neurotic	Not Neurotic	Row Total
From Broken Home	a	b	Sum a,b
Not From Broken Home	c	d	Sum c,d
Column Total	Sum a,c	Sum b,d	Grand Total

but who did come from broken homes, the members of Group **c**, those who are neurotic and didn't come from broken homes, and maybe even the members of Group **d**, those who are not neurotic and are not from broken homes.

Now suppose that sample data are gathered having to do with both neurotics and nonneurotics and that this information is as shown in Table 12-3.

One thing that stands out as we glance at this table is that only a minority of both groups came from broken homes (12 out of 30 versus 30 out of 70). Remember: These numbers are our *observed frequencies.*

It will be useful to obtain some numbers for comparison purposes based on what one would *expect* to get if there were *no relationship* between homelife and neurosis. How might we achieve that?

For Cell **a** of Table 12-3, we would do some simple calculations of the kind (Sum **a**, **c**)/Grand Total = 30/100 = .3. This is the probability that a person selected at random will be one classified as Neurotic. This figure also tells us what percent of those From Broken Homes, the 42 people in the top row, would be expected to be Neurotic if neurosis were completely unrelated to the homelife categories. That is, .3 times 42 = 12.6. This is the *expected frequency* which we could place alongside 12 in the Cell **a** position of Table 12-3. Similar calculations have been carried out for the remaining three cells.

Table 12-4 shows the observed frequency data and the expected frequency data, the latter in parentheses, side-by-side. The similarities are impressive and suggest *no relationship.* (That is, the observed number of neurotics from broken homes is about what we would expect; the observed number of neurotics not from broken homes is about what we would expect; and so forth.)

Table 12-3. Relationship Between Homelife And Mental Health (Hypothetical Data)

	Neurotic	Not Neurotic	Row Total
From Broken Home	12	30	42
Not From Broken Home	18	40	58
Column Total	30	70	100

What would the data in Table 12-3 have to look like to support suspicions of a relationship? Clearly, Cell **a** should have an observed value notably greater than 12.6; over the entire table, larger differences between observed frequencies and the corresponding expected frequencies would have to be seen. Table 12-5 shows one such possibility. This table says rather emphatically, "Yes, there is a relationship."*

Table 12-4. Observed And Expected Frequencies (Hypothetical Data)

	Neurotic	Not Neurotic	Row Total
From Broken Home	12 (12.6)	30 (29.4)	42
Not From Broken Home	18 (17.4)	40 (40.6)	58
Column Total	30	70	100

*A formal test for this kind of data is the chi-square test for two systems of classification found in most statistics textbooks.

Table 12.5. Evidence Of A Relationship (Hypothetical Data)

	Neurotic	Not Neurotic	Row Total
From Broken Home	27 (12.6)	15 (29.4)	42
Not From Broken Home	3 (17.4)	55 (40.6)	58
Column Total	30	70	100

But, as noted, the "real" table, Table 12-4, says just as emphatically, "There is no such relationship." It also says, "*Cell **a** thinking* ('One thing obviously leads to the other thing; I have seen it again and again—and don't confuse me with facts about any alleged other three cells') is wrong in this case." Let us consider another example:

Thomas Gilovich* points out that it is widely believed that couples who cannot have children should adopt. Adoption leads to conception. So says the folk wisdom.

This is another example of Cell **a** thinking (adoption obviously leads to conception), which, of course, doesn't necessarily make it wrong. But the risk of wrongness is high because only part of the relevant information is considered.

We must not forget that there are four cells in the table (Table 12-6). Only one belongs to infertile couples who adopt and then have a child of their own. When that does happen, it is imbued with high drama by family, friends, and medical staff. Still, there *are* infertile couples who adopt and subsequently do not have a child of their own. There are

**How We Know What Isn't So* (New York: The Free Press, 1991), p.3.

also infertile couples who do not adopt but do have a child of their own and infertile couples who do not adopt and do not have a child of their own. Overall, research has shown that the belief about adoption leading to fertility fails to hold up. It might be as plain as day, but it is still wrong.

Table 12-6. Relation Between Adoption And Fertility

	Child	No Child	Row Total
Adopt	a	b	Sum a,b
Do Not Adopt	c	d	Sum c,d
Column Total	Sum a,c	Sum b,d	Grand Total

NUMBERS ARE NICE, BUT ...

Can psychics really solve crimes that baffle the police? How exciting it must be to participate in an investigation in which a psychic comes up with a truly helpful mental impression! Let us think about the question in two stages. The first stage summarized in Table 12-7 and the second in Table 12-8.

Table 12-7. Partial Contingency Table For Psychic Example

	Right
Psychic's Guesses	3

Table 12-7 is admittedly a poor excuse for a contingency table because it only shows Cell **a**. But, you see, that's the point. So often in life we are only shown Cell **a** when we really need to know much more to arrive at an informed judgment.

Imagine the message of this particular Cell **a** as a newspaper headline:

GIFTED PSYCHIC MAKES THREE RIGHT GUESSES AGAINST IMPOSSIBLE ODDS!!!

We see this kind of foolishness practically every time we stand in a grocery store checkout line. Wouldn't it be good to know how many wrong guesses the "gifted psychic" made? That is Cell **b** information and you will be lucky to find it anywhere in the related article.

What about other investigators, especially the police? How many right guesses did they make? (Cell **c** information.) How many wrong guesses? (Cell **d** information.) "But," you argue, "That stuff isn't even news." News or not, all this information is important if the usefulness of psychics to police work is ever going to be evaluated properly.

I would submit that Table 12-8 is a vast improvement over Table 12-7. "But where," you ask, "are the rest of the numbers?" A grim fact of life is that often we will not have them. But that doesn't necessarily mean we shouldn't think about them! That is, think about how the rows and columns would be labeled and about how Cells **b**, **c**, and **d** should be interpreted.

Table 12-8. More Complete Contingency Table For Psychic Example

	Right	Wrong	RowTotal
Psychic's Guesses	3		
Guesses Of Others			
Column Total			

By ignoring cells, albeit imaginary cells, I recently committed an unforgivable blunder. It seems that my college-age daughter was involved in an automobile accident. That wasn't the blunder—not mine, anyway. My blunder occurred when my daughter informed me that there were no injuries except that the other driver suffered a cut lip when his airbag exploded. Having heard unflattering airbag stories before, I quickly concluded that airbags do more harm than good. I expressed this hastily assembled view to an acquaintance who responded with, "But you don't know how things might have been without the airbag." Now this man is not a Master Statistical Critic. He is not even a Fledgling Statistical Critic. But he was right.

"Why?" I wondered, "did I not have the presence of mind to envision a contingency table with No Injury as one column and Injury as the other column and Airbag and No Airbag as the rows?" Had I done so, I probably would have imagined a large observed frequency in the Airbag/No-Injury cell.

"Why," I continued, "did I let myself get trapped in Cell **a** thinking when I spend so much time telling others not to." My uncertain answer is that Cell **a** thinking is the kind of thinking we drift toward when we are not on the alert.

Now please don't misinterpret the intent of this section. Pie-in-the-sky contingency tables like those introduced here—ones existing only in the imagination and having no or few actual observed frequencies—are not scientific instruments by any stretch of the imagination. Leave them out of any future scientific papers. Leave them out of any future expert-witness testimony. Pretend you have never heard of them, if necessary, to placate scientifically oriented peers. This section is not about science. This section is about stretching the mind beyond the usual, often very comfortable, confines of Cell **a** thinking. That's not too shabby, even if it isn't science.

WHEN FALSE POSITIVES HAPPEN TO GOOD PEOPLE

Life abounds with circumstances for which Cell **a** thinking offers insufficient guidance. For example, suppose you lost your job through no fault of your own. Suppose further that you have been job hunting for over a year under conditions of progressively greater financial hardship. Finally, suppose that one morning you see advertised in the classified section of the newspaper a job opening for which you are well qualified.

But beware! At the bottom of the ad are the words: DRUG TEST REQUIRED. But who cares? You don't use the lousy stuff; you'll pass the test. It's as plain as day. (Note that the no-drugs-therefore-pass-test viewpoint is Cell **a** thinking. See how slyly it creeps in and how much sense it seems to make?) You send in your resume and, in a few days, find yourself being interviewed. It seems to go well. And, oh yes, you leave a specimen for the lab.

Several days pass. Then you get the word: "Sorry. That position has been filled." Some serious digging reveals why the job got away from you: You failed the drug test! "What? Isn't it supposed to be as plain as day that if you don't do drugs, you won't fail the test?"

The harsh fact of the matter is laid bare in Table 12-9 on the following page. This table would also be appropriate for honesty tests, lie detector tests, graphology analysis, and a lot of other imperfect things personnel people have inflicted on the rest of the world. Cell **b** in this table represents the false positive condition, that is, the applicant does not use illegal drugs, but the test says that he or she does.

You might derive some comfort from knowing that, as such screening techniques go, drug testing is reasonably reliable—a 5 percent false positive rate when the test is given under ideal laboratory conditions. However, these tests administered under typical workplace conditions can have a false positive rate as high as 14 percent, according to *The Journal of Analytic Toxicology.*

Moreover, if you are taking certain kinds of legal drugs—an asthma drug, for example—you can unwittingly make the test think you are a lawbreaker who should be turned away. And if you should ever stoop so low as to eat bagels with poppy seeds on them, you just might have to kiss your dream job good-bye, a possibility that two Navy doctors were reportedly faced with until a deeper investigation cleared their good names.*

Before leaving this subject, we should probably look at Cell **c** of Table 12-9. This is the "false negative" cell. It has to do with people who do take illegal drugs but somehow convince the test that they don't. Conceivably, one of them has that job you were hoping for.

Table 12-9. Drug Test Possibilities

	Pass Test	Not Pass Test	Row Total
Drug Free	a	b	Sum a,b
Not Drug Free	c	d	Sum c,d
Column Total	Sum a,c	Sum b,d	Grand Total

Because statistical fallacies are difficult to classify, a writer on the subject can easily end up wondering whether he has treated all worthy topics. Here is where a hedge is useful. The following chapter is my hedge.

*This anecdote and the false-positive statistics are from Martin Yate, *Knock 'Em Dead: The Ultimate Job Seeker's Handbook* (Holbrook, Massachusetts, Adams Publishing, 1995), Chapter 25.

13

SIGNIFICANT OTHERS

"The time has come," the Walrus said
"To talk of many things;
"Of shoes—and ships—and sealing wax—
"Of cabbages—and kings—"
—Lewis Carroll

There are still several kinds of statistical bloopers that don't fit conveniently under headings used in previous chapters, but which are sufficiently common and occasionally important, that they merit treatment somewhere. Hence, this Mulligan Stew chapter.

MISCONSTRUING CONDITIONAL PROBABILITIES

The following is not an authentic Abbott and Costello routine, but it almost could be. Either way, it does illustrate the concept of conditional probability.

Costello: I am interested in your book, *How To Meet The Girl of Your Dreams.* Will it really help me?

Abbott: You are certain to meet the girl of your dreams.

Costello: Wait a minute. Are you saying that if I read your book, the probability is 1.0 that I'll meet the girl of my dreams?

Abbott: That's right. The probability is 1.0—unless, of course, you happen to be short and fat.

Costello: But I thought.... what happens if I read your book and I am short and fat?

Abbott: The probability drops to .8.

Costello: That is still pretty good. So, if I read your book and am short and fat, the probability is .8 that I will meet the girl of my dreams?

Abbott: That's right. The probability is .8—unless, of course, you wear derby hats.

Costello: Well, I wear derby hats.... So what happens if I read your book, am short and fat, and wear derby hats?

Abbott: In these times, a derby hat will drop you down to a 50-50 chance.

Costello: Ouch. So if I read your book, am short and fat, and wear derby hats, the probability is .5 that I will meet the girl of my dreams?

Abbott: That's right. The probability is .5—unless, of course, you are of Italian descent.

Costello: I am of Italian descent. So what happens if I read your book, am short and fat, wear derby hats, and am of Italian descent?

Abbott: Women think Italian men are the greatest. That pushes the probability back up to .7.

Costello: So, if I read your book, am short and fat, wear derby hats, and am of Italian descent, the probability is .7 that I'll meet the girl of my dreams?

Abbott: That's right. The probability is .7—unless, of course, you know Brad Pitt personally.

Costello: It just so happens I do know Brad Pitt personally. So what happens if I read your book, am short and fat, wear derby hats, am of Italian descent, and know Brad Pitt personally:

Abbott: You are certain to meet the girl of your dreams.

This dialog between Abbott and Costello involves *conditional probabilities*—probabilities with strings attached. Obviously, the presence of a "condition" can alter a probability dramatically.

For example, you are invited to turn to Table 12-5 (page 218), a table suggesting a relationship between neurosis and homelife. As you know from reading Chapters 9 and 12, if you wished to find the probability of a randomly selected person being designated Neurotic, you would divide the Condition Present column total by the grand total, getting 30/100 = .3. This is the regular, or unconditional, probability of Neurotic for this sample.

On the other hand, the probability that the person so selected will be Neurotic, *given that he or she came from a broken home* is obtained by dividing the Cell **a** observed frequency (the one not in parentheses) by the Condition Present row total. That is, 27/42 = .64, a result more than double the unconditional probability of .3 calculated above.

Misuses of probability resulting from confusing conditional probabilities with unconditional probabilities, or vice versa, are legion. My one-size-fits-all advice for these is: pay close attention.

Usually more troublesome, and often more serious in its consequences, is the error made when one conditional probability is confused with a similar-sounding conditional probability. In abstract form, this error occurs when someone mistakes the probability of A given B with the probability of B given A. Let us take another look at Table 12-5. We see that the probability of *Neurotic given From Broken Home*—.64, as demonstrated above—is a good deal smaller than the probability of *From Broken Home given Neurotic:* Cell **a** observed frequency divided by the Condition Present column total = 27/30 = .9.

A rather recent example of confusing two similar-sounding conditional probabilities occurred during the time of the O. J. Simpson murder trial. A spokeswoman for an anti wife-abuse charity said in a television interview, "It is virtually certain that a battered wife will one day be a violently murdered wife." My guess is she meant to say, "It is virtually certain that a violently murdered wife had previously been a battered wife," admittedly, a fine distinction, considering the tragic nature of the subject.

If this example gives you a feeling of *deja vu,* or what baseball great Yogi Berra liked to call "*deja vu* all over again," it might be because of its similarity to the suicide example in Chapter 6. Yes,we are dealing with the *opportunistic construction of a percent* all over again.

Regrettably, the transition from percents to conditional probabilities is not achieved without some loss of confidence. Unlike situations calling for the analysis of known percents, conditional probability situations tend to be a little light on numbers for the statistical critic to pester. It's kind of like surfing without a board. Fortunately, with a little experience, we can learn to think some kinds of problems through pretty well. For example:

Dumb: If I picked a member of the U.S. Supreme Court at random, I'll bet the one chosen would be male.

Dumber: Oh, I don't think so. Suppose I go outside and pick a male. I don't think it's likely at all that he would be a member of the U.S. Supreme Court.

Arbitrator: Both are smarter than their labels suggest—but they are talking about different things. At this time, seven out of nine members of the U.S. Supreme Court are male. Clearly, if you were told that a person is a member of the Supreme Court, you would have to suspect he is male. But knowing that a person is male is to know a flat nothing about his profession. Still, it doesn't take the proverbial brain surgeon or rocket scientist to divide seven male Court members by a hundred million or so male citizens and figure out that this particular male probably doesn't do lunch with Chief Justice Rehnquist.

Dumb: The probability that a player on one of the NBA basketball teams is African-American is very high.

Dumber: I must disagree. I could interview African-Americans until the cows come home and never find one who plays on an NBA basketball team.

Arbitrator: Again, if Dumb and Dumber would simply agree on what it is they are discussing, they would have fewer differences. In view of the preponderance of black players in professional basketball today, the knowledge that a person plays on an NBA team would strongly suggest that he is African-American. On the other hand, knowing that a person is African-American does not (or should not) nudge us very hard in the direction of believing he is an NBA basketball player.

I am sure you get the idea by now. I also trust that you see the wisdom of paying very close attention to the exact wording of arguments involving the essence of the conditional probability concept.

THE REGRESSION FALLACY

Sir Francis Galton (1822-1911) was a man of many achievements: He was an explorer, an anthropologist, and a pioneer in eugenics. Moreover, he was the creator of the correlation techniques referred to in Chapter 11 of this book. Ironically, he was also the father of the regression fallacy.

In his study of hereditary traits, Galton pointed out the apparent regression toward the mean in the prediction of natural characteristics. He found, for example, that unusually tall men tend to have sons shorter than themselves and unusually short men tend to have sons taller than themselves. This fact suggested to Galton *regression toward mediocrity* in heights from generation to generation.

Others have followed Galton's ill-advised lead: One sometimes hears that extremely intelligent and extremely dense parents tend to have children with I.Q.'s closer to the average. It is probably true; but to read into it a sustained tendency toward mediocre intelligence, as several authors have done, is seriously wrong.

During any given year, some business firms will experience unusually high profits relative to the average of several recent years because of nonrecurring factors such as an extraordinarily high demand for a particular product, a windfall inventory gain resulting from a commodity price increase, or any of a number of other possibilities. If such firms do not fare so well the following year, it should come as no surprise. But neither should it be interpreted as a symptom of regression toward mediocrity in business, as at least one popular book eloquently argued.

The phenomenon that Galton, and others after him, observed was real enough, but the assumption of "regression toward mediocrity" is faulty and represents one form of the regression fallacy. (This is the form where one observes regression, but misinterprets it. There is another, probably more important, form where regression is present but goes unrecognized. We will address the latter form shortly.)

Today we know that whenever two variables are imperfectly correlated, extreme values of one such variable will be paired, on average, with less extreme values of the other variable. The reason for this, it is thought, is that each of the extreme values of the one variable consists of two components: A "real" component (an extremely tall father is tall partly because he has a genetic predisposition toward tallness, one he will pass along to his progeny) and a component consisting of "chance factors" (probably unknown, but conducive to even greater tallness in this father—but not in this father's progeny.)

We now arrive at the second form of the regression fallacy. We forget how commonly present regression is, work fruitlessly, and even extravagantly, to come up with causal explanations. For example, your school choir performs in a spectacular way one time, but the next time is disappointing because the choir members appear to have "slackened off." Maybe they did slacken off. Alternatively, it just could be the regression phenomenon at work. It's pretty tough to put two spectacular performances back-to-back.

Following a crime spree in a certain city, the size of the police force was increased by twenty percent. The crime rate subsequently decreased. Was it because of the increase in the police force? Probably. But it might have just looked like it. The term "crime spree," itself, suggests an abnormally large amount of crime. Might the improvement be nothing more than the regression phenomenon?

And let us not overlook the *magazine cover jinx*. Some top athletes are reluctant to appear on the cover of *Sports Illustrated* because they have seen too many of their colleagues drop into relative obscurity after being so honored. Moreover, some followers of the stock market look with caution upon a *Time* Magazine cover lauding recent stock market performance.

Neither is sheer nonsense. Nor does either require a superstition-based explanation. The regression effect alone would suggest that an athlete, asked to adorn the cover of *Sports Illustrated*, is likely to be having a banner season—a probable portent of something less spectacular next season. Similarly, if the recent performance of the stock market has rendered it worthy of a celebration on the cover of *Time,* the bull market just might be getting pretty long in the tooth.

Let us anticipate the next section by identifying another gremlin that follows the regression phenomenon around. Studies have shown that people in general fail to take regression adequately into account when making projections. For example, if a firm that has enjoyed a 15 percent profit margin for quite awhile is now basking in the warmth of a 45 percent profit margin, it probably will do less well next year. But, chances are management will overweight the current exceptional experience when preparing its projections.

Be mindful of the regression phenomenon when evaluating projections. Typical performances tend to be followed by typical performances; exceptional performances tend to be followed by performances of a less exceptional nature.

UNCRITICAL PROJECTION OF TRENDS

An activity in which we all engage, sometimes in ways so subtle it doesn't seem like that is what we are doing, is the projection of trends. It is an essential part of planning for the future. When used in conjunction with much knowledge about the variables playing upon a series of past numbers, trend projection can be most beneficial; the use of the tool is not by itself a statistical fallacy. It is when good sense is abandoned and *arithmancy* allowed to rule that the method runs amuck.

Sometimes even the arithmancy (a word suggesting a kind of magical manipulation of past numbers) is pretty simplistic, consisting as it often does of taking **n** numbers and determining what the **n+1** value would have to be to keep the series growing in exactly the same manner as in the past, and then determining what the **n+2** value must be to effect this same end, and so forth. The process can (but should not) be continued indefinitely.

For example, if the United States had a population of 90 million in 1910 and 180 million in 1960, the arithmanticist would say that population doubles every fifty years and that we will have 360 million people in the year 2010, 720 million by 2060, and just under a billion-and-a-half by 2110. "Eventually, using this reasoning, there will come a day when the radius of human flesh will expand at the speed of light," as one expert testified before a congressional committee.

Other exciting conclusions can be reached by projecting trends unhampered by logical considerations. For example, personal computers will eventually become so small that man will have to evolve invisibly skinny fingers in order to operate them. The earth will one day be covered by concrete. And a sprinter will run the 100-yard dash in no time at all—or even finish the race before it begins.

BLIND ACCEPTANCE OF COMPUTER RESULTS

So useful has the computer become in practically all aspects of our lives that there may be a tendency for us to let it to do our thinking for us. That, generally speaking, is not a good idea. The computer cannot work magic—not yet anyway. It will do only what it is instructed to do, and the validity of the results depends greatly upon the accuracy and adequacy of the information put in and the wisdom of the people who will act upon the results. The statistical critic, of all people, should resist being overawed by the news that such-and-such information was turned out by a computer. The fact that computers are used today even for casting astrological horoscopes speaks volumes about the inadvisability of blind acceptance.

OMISSION OF STRATEGIC DETAILS

The year was 1997, the place was the University of California at San Diego, the occasion was graduation, and the speaker was none other than President Clinton himself.On that grand occasion, the President made a statement which illustrates quite nicely the message of this subsection. He said that by the year 2050, when the grandchildren of the present graduates are graduating, there will be no majority race in America. Was the statement true? Yes, you could say it was true. Well, was the statement false? Yes, you could say it was false.

As the courts have long recognized, what is needed is (1) the truth, (2) the whole truth, and (3) nothing but the truth. Sometimes statistical evidence, not to mention a president's remarks, score well on (1) and (3) but poorly on (2). Such a profile often means that something important has been withheld.

For the country to be majority-race free by 2050, the President had to resort to an unstated operational definition. It seems that the U.S. Bureau of the Census does not classify Hispanic as a separate, identifiable race. Over 90 percent of such citizens are counted as white. Moreover, this component of the white population is growing sufficiently rapidly to ensure a still-white majority by 2050. Statistical clarity would have required the President to say something like, "By 2050 non-Hispanic whites will no longer dominate the population totals." Tact, on the other hand, dictated that the President avoid use of the term "non-Hispanic white" on the grounds that it is disliked by many Hispanics. The President, ever tactful, withheld a key fact and in so doing told a little lie.*

Sometimes the difference between being enlightened by statistical evidence, on the one hand, and being misled by the very same evidence on the other, is solely a matter of something outrageously subtle—what needed to be said but wasn't.

*Related in Dan Seligman, "Lies, Damned Lies and Politically Motivated Statistics," *Forbes,* July 28, 1997, pp. 52-53.

For a number of thought-provoking, serious examples of how considerably our view of a subject can change when we step from having a few facts and a short time horizon to more facts and a longer time horizon, see Chapter One of *The Healing Brain.** The book leads us to wonder whether very many of the things we think we know about the march of medical progress over the centuries are really as we believe. Consider, for example, what we think we know about the epidemic disease tuberculosis.

> Most people believe the decline of tuberculosis was due to the introduction of the antibiotic [streptomycin]. But our perspective is wrong, our time frame is wrong, and our conclusion is wrong as well. Most people are, in effect, looking at a few exciting and vivid happenings within their lifetime, the past forty to fifty years, basing their conclusions on this familiar but very restricted evidence. A longer view reveals that most of the total decline of deaths from tuberculosis since the 1800's occurred before streptomycin was first introduced. By 1945, 97 percent of the cases had already been eliminated, leaving only the remaining 3 percent to be improved by the modern treatments. The dramatic emptying of the hospitals that many of us and our parents have observed, then, was only the very end of the massive improvement.

Tuberculosis is not the only epidemic disease, according to Ornstein and Sobel, for which excessive credit is given to medical discoveries. Others include pneumonia, influenza,

*Robert Ornstein and David Sobel (New York: Simon and Schuster, 1987).

whooping cough, measles, and scarlet fever. Other factors that were helping rid the world of these epidemic diseases well before the much-heralded medical breakthroughs were advances in agriculture, the purification of water, improved sewage disposal, the pasteurization of milk, and so forth.

ILL-CONCEIVED RATIOS

Sometimes ratios that are perfectly accurate from an arithmetic standpoint create bizarre impressions and are, therefore, worthy of brief treatment here. Someone might say to you, for example, "Since about four million babies were born in this country last year and since the total population was some 250 million or so, we must conclude that every man, woman, and child in the United States had 1/62 of a baby last year." Here are some additional examples:

- You can multiply the number of felons in prison by 365, and then average this out among the some 250 million people in the United States and discover that the average person serves about ten hours a year for a felony.

- During the sixties, arguments like the following were occasionally heard: Last year 700 million marijuana cigarettes were smoked in the United States—almost four for each man, woman, and child—in a disregard for the law not seen since prohibition.

- It is estimated that in the United States the average man buys only one-third of a pair of pajamas per year.

Undoubtedly some common statistical fallacies have been overlooked. Still, we have now dealt with quite a long list of these hindrances to straight thinking. If you can recognize the ones we have covered when you encounter them in books, speeches, newspaper and magazine articles, and TV information shows, you can quite properly consider yourself a pretty sharp guy or gal—maybe even a future Master Statistical Critic.

What's that? You say you are not sure whether that is a description of you? You're in luck. The following chapter will give you a chance to find out.

14

FORTY AWFUL EXAMPLES

> There is no merit where there is no trial; and till experience stamps the mark of strength, cowards may pass for heroes... .
> —A. Hill

This is a hands-on chapter. Follow the rules and it should prove a highly instructive one. On the following pages you will find forty "awful examples" of statistical material. Your job is to tell what makes each such example "awful."

First, though, you should know you have help. In the glossary beginning on page 260, you will find many of the statistical fallacies you have encountered in this book consolidated under just four broad headings. Definitions and text page numbers are also shown.

Fortunately, when the four broad headings are expressed in question form, the result is a simple, easy-to-remember guide to evaluating most kinds of everyday statistical information. The four questions are:

1. **Is there anything about the statistical information or its presentation that just doesn't look right?**
2. **Is something being confused with something else?**
3. **Are there signs of opportunistic selection?**
4. **Has anything strategic been withheld or overlooked?**

Lucky you when you encounter a **Question 1** kind of "awful example." A checklist of fallacies based on Part I of the glossary is probably all you will need. Do any charts have unmarked axes? Are there any dangling comparisons? Any hyperaccuracy? Any meaningless statistics?

Most "awful examples" of the **Question 2** kind will be similarly clear cut. With these, you will be asking yourself, "Did the source say 'percent change' when he or she should have said 'percent points of change'?" "Is the source reading causation into a mere correlation?" And so forth.

With "awful examples" related to **Question 3**, you will be stepping into deeper water. You will find yourself asking, "Why did the source use the arithmetic mean rather than the median?" "Why did the source use that particular base value when there are so many other possibilities.?" And so forth.

With "awful examples" related to **Question 4,** the water is deeper yet. With these, you will be worrying about why the source failed to include a certain kind of information. "Did the source carelessly overlook or intentionally withhold something important?" These will require much thought.

On with the challenge. My answers begin on page 252. But for any specific example, don't look at my answer until you have given it a serious try. Also, expect some differences of opinion.

1. A marketing research study uncovered this fact: The average age at which young men in Akron, Ohio, begin shaving is 15.5371 years.

2. A sorority dance was considered a huge success by its planners because only four people out of the 200 in attendance complained. "When only four people are discontent and 196 are delighted," one of the planners was heard to say, "as far as I'm concerned, that's a successful dance!"

3. A television commercial for *Investor's Business Daily* claims that the newspaper's readers have the highest income and net worth of any American business newspaper or weekly business magazine. Then it asks why people who can afford any business-related publication read *Investor's Business Daily.* The answer, according to the commercial, is that *Investor's Business Daily* is the only business newspaper with "exclusive" information for those with an interest in making big money.

4. An employee of a cellular phone company was about to be transferred to another state and had his choice between Kentucky and California. Since climate was a matter of considerable importance to him, he did some investigating into the temperatures of the two states. He learned that, over the course of a year, the two states have nearly the same average temperature. He concluded that insofar as temperature was a factor, it didn't make any difference to which state he was transferred.

5. The facts presented below are all from the same newspaper article; the heading was, **WOMEN OUTDRIVE MEN !**

(1) Of men involved in accidents, 23% had been drinking—some 9.6% for women.

(2) One third of the men were driving too fast, one fourth of the women.

(3) Of 101 drivers involved in an accident while passing on a curve, 15 were women.

6. Between 1946 and 1970, the total amount of money spent on residential structures (essentially houses and apartments) in the United States increased by more than fourfold, as shown in Figure 14-1.

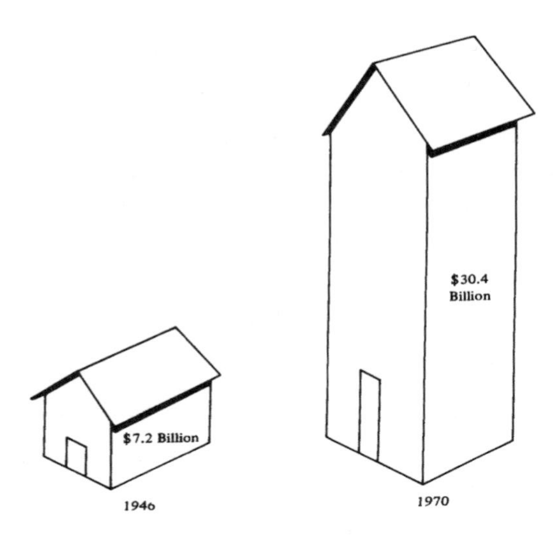

FIGURE 14-1

7. Bernie Skiles hit two home runs his first two times at bat during an important game. As he approached the plate for the third time, a spectator was heard to say, "Old Bernie won't get a home run this time; the odds against three home runs in a row are astronomical."

8. A study found that, although the severe poverty group and the comfortable groups differ in income, family structure, etc., they are quite similar in the value they attach to educational goals. The study was conducted by questioning a random sample of all students in 11 colleges in the deep South.

9. In a newspaper article attempting to show that women drive better than men, it is stated that women even drive bicycles better: "Going further, the patrol looked into bicycle accidents. Some 3000 males were injured on bicycles in the state last year and 34 were killed compared with 662 females injured and 11 killed."

10. A manufacturer registered the following complaint with a government agency: "We have been hard hit by material allocations. As a matter of cold fact, there has been a 150 percent decline in our production in this quarter compared to the same quarter last year. You know what that does to costs."

11. A certain company experienced a 23 percent increase in its per-unit cost of production and only a 15 percent increase in per-unit selling price during a certain five-year period. The company's chief negotiator lamented to the head of the employees' union that this 8 percent decrease in per-unit profits was putting the very future of the company in jeopardy. Consequently, he urged the union to lower its wage demands.

12. A man's income increased from $4000 during his last year in night school to $8000 the following year—an increase, he figured, of 50 percent.

13. Peter, Paul, and Mary got appreciably higher scores on the final examination for a certain course than they had achieved on the midterm examination. Peter's score increased by 25 percent, Paul's by 20 percent, and Mary's by 15 percent. Peter asserted that, among them, they had raised their scores by 60 percent: (25+20+125) = 60%. Paul argued that the correct percent was 18.3: (25+20+15) divided by 3 = 18.3. Mary laughed.

14. A salesman for *Investor's Business Daily* told me over the telephone, "Eighty-two percent of the people who switch from the *Wall Street Journal* to *Investor's Business Daily* don't even look at the *Wall Street Journal* after that."

15. A teacher noted that many students who had low scores on the midterm examination were much closer to the class average on the final. This she attributed to her teaching skill. But she also noticed that several students who were exceptionally high on the midterm examination slumped noticeably on the final. This she attributed to slackening off because of overconfidence.

16. A television panel consisting of professional comedians got into a discussion about the connection between being a successful comedian and having been the "class clown" in school. All of the panel members admitted to being former class clowns. More impressive yet was the fact that the names of other top comedians were brought up and, it was revealed, nearly all had been class clowns. The moderator then looked straight into the camera and said very seriously, "There you have it. If you are the class clown, the chances are excellent that you will one day be a famous comedian."

17. It was found that 14 percent of the divorce cases in a certain country involve men who had married between the ages of 25 and 30 and that only 11 percent involve men married between the ages of 30 and 35. These figures were taken as proof that the older a man is when he marries, the more likely the marriage will be successful.

18. A newspaper article claimed, "Graduates can afford to be choosy about careers these days. For one thing, there are so many more occupations—21,741 at the last count by the Department of Labor."

19. A magazine article stated, "There are 48 million Americans who call themselves Roman Catholics... .from a nationwide survey of U.S. Catholics above the age of 17, ..., a strikingly bleak picture emerges. More than a third do not attend Mass regularly." This information was interpreted as an indication that many Catholics are falling away from their religion.

20. It was determined from a reputable study that 75 percent of the people using Sparkle-Plenty Toothpaste for a trial period developed no cavities. This finding was written up in a newspaper article entitled "Nonusers of Sparkle-Plenty get cavities."

21. A recent divorcee comforted himself by saying, "My next marriage is bound to work. The odds against two divorces are huge."

22. Marketing researchers for a certain automobile company prepared a questionnaire for Panther owners (a small sports car manufactured by this company) to determine the characteristics these owners prefer in future Panther models. The questionnaire was completed by all participants of a Panther Club national convention. The results of this survey were taken into account when future models were designed.

23. A newspaper article asserted that children do not copy their parents' smoking habits. According to a survey including some 50,000 children aged 11 to 18, parents' smoking habits had almost nothing to do with whether the children smoked. Among smokers, 70 percent had fathers who smoked; but 63 percent of the nonsmokers also had smoking fathers.

24. A magazine article asserted that 75 to 80 percent of all convicted criminals are products of broken homes and unhappy childhoods.

25. A certain football team has six members who weigh 240 pounds, eight who weigh 225 pounds, three who weigh 180 pounds. Their arithmetic mean weight, as reported by the coach, is 211.25 pounds.

26. A certain company has 1,900,000 shares of stock outstanding and 3,700 common stock holders, two of whom own 1,000,000 shares between them. In the annual report, the statement is made: "Our common stock holders own an average of almost 514 shares apiece."

27. A survey conducted by a large national personnel agency finds that discrimination against the overweight is rampant in business. Fat people, it says, are woefully underpaid; "the overweight have become America's largest, least protected minority group."

28. It is imperative that a certain kind of plastic tubing be no less than 12 inches in length. A quality control engineer examines a particular batch and finds that the average length is 14 inches. He concludes that the batch is satisfactory.

29. An advertisement began "One out of three people you work alongside is a law breaker of some sort."

30. Between 1946 and 1970 the total amount of money spent on residential structures (essentially houses and apartments) in the United States increased by more than fourfold, as shown in Figure 14-2.

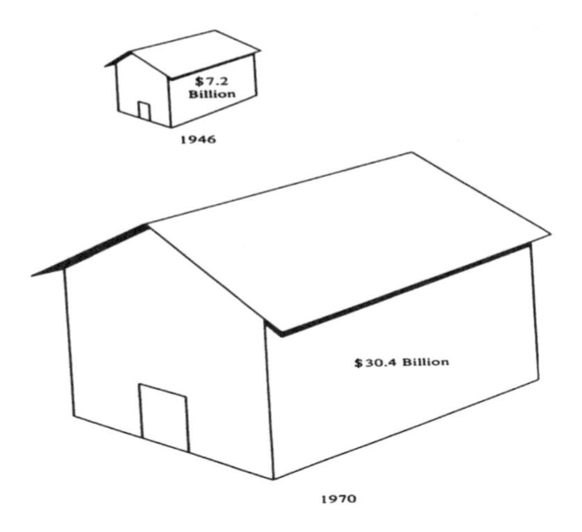

FIGURE 14-2

31. An article entitled "They Put a Parson on the Payroll" in a popular magazine states: "In just two years, religion-on-the-job has accomplished several pretty wonderful things ... labor turnover has dropped from 7.61 to 5.22% in two years, the accident rate has declined approximately 40% and absenteeism is much lower than it used to be."

32. **Moe:** Only 20 percent of dog owners buy flea collars.

Larry: I don't think that's right. I read just the other day that over 90 percent of the people who buy flea collars own dogs.

Curly: I knew a flea caller once. She could yell the fleas right off the dogs at 20 feet.

33. An advertisement for a driving school boasts that, whereas the Harvard Law School graduates 1 out of 6 applicants, this particular driving school graduates only 1 out of 10 applicants.

34. There are more than 6,000 closet-transvestite lawyers in the United States.

35. Here is some wry statistical humor, compliments of MarkTwain's *Life On The Mississippi* :

In the space of one hundred and seventy-six years the Lower Mississippi has shortened itself two hundred and forty-two miles. That is an average of a trifle over one mile and a third per year. Therefore, any calm person, who is not blind or idiotic, can see that in the Old Oolitic Silurian Period, just a million years ago next November, the Lower Mississippi River was upward to one million three hundred thousand miles long, and stuck out over the Gulf of Mexico like a fishing-rod. And by the same token, any person can see that seven hundred forty-two years from now the Lower Mississippi will be only a mile and three-quarters long, and Cairo and New Orleans will have joined their streets together and be plodding along under a single mayor and a mutual board of aldermen. There is something fascinating about science. One gets such wholesale returns of conjecture out of such a trifling investment of fact.

36. The participation rate for a local charity drive went from 45 percent in 1995 to 65 percent in 1996. In other words, it increased by 20 percent.

37. The typical high school male dates two and three-quarters girls a year.

38. In a television commercial, a man with a microphone runs out into the street and asks people in each of three haphazardly selected cars (apparently waiting for the light to change) whether he or she uses a certain dental product. Two answer "yes" and one answers "no."

39. A public relations woman for the airline industry feels an obligation to make air travel seem safe relative to other forms of travel. She states, truthfully, that the annual number of automobile fatalities has increased almost every year since the end of World War II and is right now at an all-time high.

40. A news magazine reported the following experience:

Most doctors know that visitors often do more to stir up hospital patients than to soothe them. But the doctor's own ward rounds can have the same effect, sometimes with fatal results, reported Finnish Doctor Claus Jarvinen in the *British Medical Journal.*

Studying the histories of 39 Helsinki hospital patients who died of coronary occlusion after stays of seven to 42 days, Dr.Jarvinen discovered that six of them, subject to severe emotional stress, had died during or after a physician's visit. Among the cases:

An accountant, 58, came to the hospital 21 days after an attack of angina pectoris. He seemed in satisfactory condition until the 16th day in the hospital. The head physician was making his round; as the doctor drew closer, the patient became nauseated, suffered a severe attack and died within two hours.

After suffering chest pains during a tantrum, a female post-office clerk, 68, was admitted for treatment. In the ward, she grew excited over trivialities. After nine days, when the doctor approached she became restless. Asked how she felt, she tried to answer, and died on the spot ... *.

***Time,* February 21, 1955, p. 37. Reprinted by permission.

AUTHOR'S ANSWERS

1. Sounds like Hyperaccuracy.

2. Using the number complaining as a proxy for the number discontent greatly underestimates the truth.

3. Although this isn't easy to classify, I would call it a rather elaborate meaningless statistic. Sometimes a statistical argument will lead a none-too-attentive reader or listener to form an impression completely unrelated to the statistical facts given. Although this commercial dances around the thought that if you read this newspaper, you will get rich, never is such a connection actually made.

4. The averages might be comparable, but the standard deviations differ considerably, the larger one being associated with Kentucky. Kentucky's temperatures would do more swinging between uncomfortably hot and uncomfortably cold days.

5. All three points fail to support the claim that "Women Outdrive Men." Remember that all the drivers referred to have been in accidents, not a very good set for determining which gender drives better. All three statements have to be interpreted as," 23 percent of men had been drinking, given that they had been involved in an accident; 9.3 percent of the women had been drinking, given that they had been involved in an accident;" and so forth. This appears to be a conditional probability problem or an opportunistic construction of a percent. Whatever one calls it, the claim made is simply not supported by the evidence given. At best, this suggests that certain driving practices may differ between the sexes.

6. The height of the right-hand house looks accurate, but it is a strange looking house. The temptation is probably great to widen it, something that could introduce serious technical problems. Even as is, the pictures are of three-dimensional objects; it is questionable whether the viewer would interpret the chart as intended. A columns chart would be better.

7. Assuming independence, probably a pretty safe assumption, the probability of hitting that third home run is the same as it is for hitting any home run. Since the first two home runs are now fact, there is no uncertainty about them.

8. Of course, much depends on the purpose of the study and the definition of the population. However, one can rightfully wonder what the researchers are seeking. All the respondents are already in college; certainly they will share views about educational goals. Moreover, the students not questioned are presumably mostly in the severe poverty group. Offhand, it seems as if there is a huge bias in the sample.

9. First, one would be justified in asking whether bicycle statistics have any bearing on how the people under scrutiny drive automobiles. Beyond that, it's hard to know what we are supposed to be impressed by. Men have many more bicycle accidents, but that could just be the fallacy of the sheep. If being killed on a bicycle is what this is about, women are seen to have a higher percentage.

10. For something that can only have positive values, the largest decline possible is 100 percent.

11. This is not the way to figure percent change in per-unit profits. Suppose, for example, cost per unit had been $1.00 and selling price per unit had been $3.00. This means that the former goes to $1.23 and the latter to $3.45, a difference of $2.22. Thus, profit per unit used to be $3.00 - $1.00 = $2.00 and is now $2.22. This is an increase of 11 percent rather than a decrease of 8 percent. This kind of problem requires knowledge of the base values.

12. 100 percent. [($8000-$4000)/$4000)]•100 = 100%.

13. One must have knowledge of the midterm scores in order to work this correctly.

14. I'm guessing that this is one of those miraculous statistical births dealt with in Chapter 2. If not, I think we can still charge it with at least one count of meaninglessness, in view of the vagueness of "don't even look at ... ."

15. The teacher might be right on either or both counts. However, since she is focusing on extremes of both kinds, it seems possible that she is observing the regression phenomenon at work.

16. The fact that a high percentage of famous comedians had been class clowns says nothing about what percentage of class clowns become famous comedians.

17. The evidence offered here seems pretty meager for such a huge inductive leap. First, there are only two age categories; very young and very old married men are not represented. Second, the two percents, 14 and 11, are sufficiently close that it might not pay to make too much out of the difference.

18. It isn't at all clear that the information given supports the conclusion drawn. A longer list of occupations hardly makes finding a job any easier—let alone give the job seeker an opportunity to be choosy. Moreover, it would be useful to know how many occupations were on the list in the past when hiring conditions were good/bad.

19. The conclusion might be right. However, "falling away" cries out for some kind of comparison. If some research turned up that in 1950, say, half of the Catholics in the United States did not attend Mass regularly, some celebrating might now be in order.

20. This is clearly a case of using a conditional probability to draw a conclusion about a totally different conditional probability. The fact that a high percentage of users gets no cavities tells us nothing about what happens to nonusers.

21. It is possible that consecutive divorces are not independent events. Nevertheless, this man is obviously thinking that since he has already had one divorce (a certainty), the probability of a second divorce will be dramatically lowered. It ain't necessarily so.

22. This sounds like a convenience sample. Its validity, or lack of it, depends importantly on what the relevant population is perceived to be. One thing is clear, however, the people in attendance at a Panther Club national convention are probably much more aware of, and perhaps more opinionated about, Panthers, and maybe cars in general, than Panther owners who do not attend the conventions and non-Panther owners. The preferences of

this avid group might not be the same as for the other groups. The sample is certainly biased; how important that bias is is unclear.

23. There is one obvious discrepancy between what the numbers represent and what conclusion is drawn: either mothers should be part of the study, or the conclusion be made more humble, referring only to fathers. That aside, I also see this as a case of Jumping To Conclusions. Although the difference between the two percents does not suggest a black-and- white difference between the two groups, with a sample size of 50,000, the difference between 70 percent and 63 percent might very well be meaningful.

24. This needs much augmentation. First, operational definitions are needed for "broken homes" and, especially, "unhappy childhoods." Second, we need a similar percentage for noncriminals.

25. Two hundred and nineteen pounds.

26. The statement is probably true enough. However, the "average" used is obviously the arithmetic mean. In view of the skew data, the median would probably be more representative.

27. Maybe. But let us not rule out the possibility that the study might have cause and effect reversed. Maybe the overweight tend to be slower and less self-disciplined than thinner people. Maybe those characteristics get translated into lower wages or salaries. I'm guessing—maybe very incorrectly. Maybe very unpopularly. But trying out cause as effect and effect as cause is not a bad practice in general.

28. In view of the stated requirement, more attention should be paid to variation. If the standard deviation is, say, 4 inches, then probably several tubes in the batch are not long enough.

29. The statistic means nothing without an operational definition of "Lawbreaker."

30. Although its height is accurate, the house on the right is vastly larger than it should be to depict a fourfold increase. A columns chart would be better.

31. As always, one must be cautious about embracing alleged causes. The parson might have contributed to these favorable changes, but other contributing factors might also have been at work. For example, in view of the high employee turnover rate, one can hardly help wondering whether some changes for the better weren't about due anyway (the regression effect?). Moreover, although one can perhaps believe that the parson helped the turnover and absenteeism rates, it is more of a stretch to believe he helped lower the accident rate.

32. Moe and Larry might both be right. The fact that there is a conflict, however, suggests they don't realize they are reciting percents derived from different bases. Curly is equivocating—or something.

33. Well, it is certainly a comparison of unlike things. Shall we settle for that?

34. An Unknowable Statistic.

35. It speaks for itself: uncritical projection of a trend.

36. There appears to be some confusion between percent change and percent points of change. The 20 mentioned here is the percent points of change. The percent change is presumably (20/45)•100 = 44.4 percent

37. The statement might be true, but expressed in this manner, it qualifies as an Ill-Conceived Ratio.

38. The viewer, unless on guard, not only tucks away the impression that two out of every three people use this product but also that the sample is random. The interviewer picks out three cars quite arbitrarily, hence, the seeming similarity to random sampling. But this is still a convenience sample and not a random one.

39. We are told that this is a true statement. However, relevant information is being withheld, namely that the number of automobile fatalities relative to cars in use or relative to miles traveled by car has been trending downward.

40. For openers, the precise meaning of "after a physician's visit" must be clarified by an operational definition. More importantly, we need to realize that only Cell **a** (Re: Chapter 12) is being examined. In view of the data we know to be available, it seems reasonable to believe that related data are also available. It would at least be desirable to know how many patients died in the hospital but not in close proximity with a doctor's visit and how many patients did

not die despite long stays and several visits by the attending physicians. We might still conclude that having a doctor visit is a risk factor, as does the author of the journal article referred to above. But then again we might not. We certainly won't know from the tiny amount of anecdotal information we have been handed.

THE END

GLOSSARY

Statistical Fallacies Arranged By Broad Categories

I. Statistical Fallacies Resulting From Simple Carelessness Or Inept Presentation

Carelessness in computing a percent. A percent is reported but is wrong because the source paid insufficient attention to the details of calculation. *See pages 110-112.*

Chart with unmarked axes. Such a chart gives a visual impression of something, but withholds the very information needed to convey either magnitudes or rates of change. *See pages 73-74.*

Combining percents. Common arithmetic operations (addition, subtraction, averaging) are performed on percents unthinkingly with erroneous results. *See pages 113-114.*

Dangling comparison. Explanatory material makes the statistical information sound like part of a comparison, but there is no other information to compare with. *See page 134.*

Hyperaccuracy. Statistical information has the appearance of a very high degree of accuracy. Usually, information about the actual crudeness of the basic data or methodology is omitted. *See pages 47-48.*

Ill-conceived ratio. A ratio is stated in such a way as to encourage a bizarre mental impression. *See pages 239-240.*

Meaningless statistic. A precise statistical fact is given in connection with a term whose meaning is not obvious and no operational definition is given. *See pages 21, 32, and 45-46,*

Pictogram. A graphic comparison is subtly exaggerated through use of pictures of normally three-dimensional objects rather than less interesting, but mathematically better, bars. *See pages 75-79.*

Uncritical projection of a trend. Future values of a variable are estimated by the mechanical application of a formula, unaccompanied by information about other relevant variables or by plausibility checks. *See pages 234-235.*

Unknowable statistic. A statistical fact is presented about an activity for which it is effectively impossible to obtain accurate data. The methodology employed is usually kept vague. *See pages 32-33, 51-53.*

II. Statistical Fallacies Resulting From Confusing Something With Something Else

Coincidence confused with a true low-probability event. Probabilistic reasoning is applied improperly to a coincidence with a result that seems remarkable. *See page 22, 34-35, and 164-167.*

Conditional probability confused with another conditional probability that sounds similar. The verbal equivalent of P(A|B) = P(B|A), where P(A|B) is "the probability of A given B" and P(B|A) is "the probability of B given A." *See pages 225-231.*

Contributory cause thought to be the sole cause. Usually an effect is the result of more than one cause. This fallacy overlooks that fact. *See pages 199-201.*

Correlation taken as evidence of causation. Many things correlate. But the presence of correlation does not necessarily indicate that the variables under scrutiny are causally related. *See pages 201-210.*

information from a self-selected sample confused with information from a representative sample. Self-selected samples run a high risk of bias. They should not be accepted casually as representative samples. *See pages 185-189.*

Joint probability after one outcome is known confused with a joint probability when no outcomes are known. The verbal equivalent of mistaking 1 times P(B) with P(A) times P(B). *See pages 161-164.*

More than one meaning used in connection with a term or measurement (equivocation) A logical fallacy. It has statistical relevance when one refers early on to a statistical measurement in a manner consistent with the source's operational definition and later refers to it in a way suggesting a very different definition. *See page 136.*

Nonprobability sampling confused with probability sampling. Information from a nonprobability sample is presented (or is interpreted) as if from a probability sample. *See pages 183-184.*

Percent change confused with percent points of change. The difference between two percents is interpreted as if it is a percent change (or percent difference). *See pages 21, 34, 107-111.*

Temporal priority confused with causality (the post hoc fallacy). Reasoning of the kind: "If A preceded B, then A must be the cause of B." *See pages 194-196.*

III. Statistical Fallacies Resulting From Opportunistic Selection

Base period of a percent is unrepresentative. The denominator used in calculating a percent or a series of percents is painstakingly selected so as to support someone's vested interest. *See pages 116-119.*

Base value of a percent is extremely small. The dramatic impact of an argument is enhanced by the presence of a large percent—alas, largely a mere artifact of using a tiny base value *See pages 115-116.*

Broad-base fallacy. Someone has possession of a percent which potentially can support his/her vested interest. However, because that percent is quite small, he/she decides to render it more impressive by multiplying it by the number of people (or items) making up a very large, more or less relevant population. *See pages 127-128.*

Opportunistic construction of a percent. Person A can benefit from a large percent; thus, he/she uses something small for the denominator. Person B can benefit from a small percent; thus, he/she uses something large for the denominator. Although the subject under contention is the same for both, they present dramatically different statistical evidence. See pages 20, 24-26, 28, 31, and 119-126.

Using the arithmetic mean as a measure of central tendency when the median would be more honest. Someone can benefit from presenting a very high (or low) representative value. Thus, he/she boasts about the high (or low) arithmetic mean even though the measure is greatly influenced by a few very high (or low) values. Under such circumstances, the median, or possibly the mode, would be more representative of the overall data set. *See pages 84-89.*

IV. Statistical Fallacies Resulting From The Withholding/Overlooking Of Important Information.

Absence of a comparison when one is desirable. The information given lacks meaning without other information to compare it with—but none is provided. *See pages 141-144.*

Blind acceptance of computer results. Statistical information is thought to be above criticism because it was turned out by a computer. *See pages 235-236.*

Cell a thinking. This is a term dependent on a 2X2 contingency table format. It entails concluding that the phenomenon associated with the Condition Present column is obviously the result of the phenomenon associated with the Condition Present row. Cells b, c, and d are not considered. *See pages 215-222.*

Disregarding the regression phenomenon when formulating an explanation. This can occur when no heed is paid the very common tendency for extreme events to be followed by events of a less extreme nature. A fruitless search for a causal explanation often ensues. *See pages 231-234.*

Fallacy of the sheep. Named for the story about research into why white sheep eat more than black sheep. The reason: There are more white sheep. More generally: When one group surpasses another with respect to some quantitative standard, don't overlook the possibility that the winning group might be larger. *See pages 137-140 and 147-148.*

Ignoring changes in definition The data supplier's operational definition changes over time. The change is reported but not noted by the user who makes an incorrect interpretation as a result. *See pages 134-135.*

Ignoring changes in reporting methods. Similar to the preceding one. *See page 135.*

Miraculous statistical birth. A metaphor for a made-up statistic, one whose shady origin is not made known. *See pages 13, 21, 32-33, and 50-54.*

Obliviousness to an equally plausible alternative conclusion. This entails jumping to a conclusion—one probably less accurate than one would arrive at after more reflection. *See pages 148-152.*

Obliviousness to a lurking variable. The data user is determined to make sense out of a scatter-diagram that seems to make no sense. In truth, both variables in the study are seriously influenced by another variable outside the study. *See pages 206-209.*

Obliviousness to variation. An analysis is limited to measures of central tendency even though variation is also important. *See pages 99-104.*

Omission of a strategic detail. An improper interpretation is encouraged because of the withholding of an essential piece of information. *See pages 136-137 and 236-239.*

Risk your life and live longer fallacy. Because fewer deaths [accidents, or whatever] occur under one condition than under an alternative condition, the first condition ensures greater safety in the future. There is usually a concealed factor. *See pages 19, 22-24, and 140-141.*

Sampling methods not revealed. Information obtained from sampling is presented as if there were no need for further elaboration, even though the sampling procedures used are important to know. *See pages 172-173, and 189.*

Undeclared proxy. A related measure is used in lieu of the more appropriate measure, but the substitution is not announced. *See pages 48-50.*

INDEX

D

E

F

G

H

I

J

K

L

M

N

O

S

T

U

V

W

X

Y

Z

Author **Steve Campbell** is Professor Emeritus at the University of Denver where he taught statistics for over thirty years. He earned his Ph.D. at Columbia University and has authored two previous books, *Flaws and Fallacies In Statistical Thinking* (Prentice-Hall) and *Applied Business Statistics* (Harper & Row). He has often served as consultant and expert witness on matters statistical. Steve recently founded Think Twice Publishing, an audacious little company dedicated to the proposition that critical thinking can be made fun. He has a wife, Judy, and two children, Scott and Melanie.

Artist **Mark V. Hall** has been a professional cartoonist-caricaturist for over seventeen years. His company, managed by his efficient wife, Jana, is called Caricature Art, Inc. and is located in Littleton, Colorado.

In addition to his many caricatures, Mark has done cover art for three previous books, *Radio Report, On The Road With Bob Hatch,* and *A Taste of Friendship.*

Mark V. Hall